Also by Lyn Hancock, with David Hancock

ADVENTURES WITH EAGLES

WILD ISLANDS

There's a Seal

in My Sleeping Bag

There's a Seal

in My Sleeping Bag

LYN HANCOCK

A L F R E D · A · K N O P F / *New York*

1972

THIS IS A BORZOI BOOK
PUBLISHED BY ALFRED A. KNOPF, INC.

First American Edition
Copyright © 1972 by David and Lyn Hancock

All rights reserved under International and Pan-American
Copyright Conventions. Published in the United States
by Alfred A. Knopf, Inc., New York. Distributed by
Random House, Inc., New York. Originally published in
Great Britain in slightly different form by Collins
Publishers, London. Copyright © 1972 by
David and Lyn Hancock

Portions of this book have previously appeared
in *Weekend* Magazine, the *Australian Women's*
Weekly, Islander, and the Victoria *Daily Times.*

Library of Congress Cataloging in Publication Data
Hancock, Lyn.
    There's a seal in my sleeping bag.
    1. Zoology—British Columbia—Popular works.
    2. Seals—Legends and stories. I. Title.
    QL50.H315   591.9'711   70-171115

    ISBN 0-394-48032-5

Manufactured in the United States of America

To a Sea Lion Called S A M

Contents

1 · A Date to Count Eagles | 3

2 · A Seal on My Honeymoon | 16

3 · Nine Hours in an Eagle's Nest | 34

4 · Sam Goes to Jail | 61

5 · Sea Parrots on Solander | 75

6 · Mr. Hancock, I Presume? | 115

7 · Mother to a Murre | 142

8 · A Home at Last | 180

9 · Killer Whales and Eagles | 208

10 · South to San Miguel | 251

11 · The Morning of Another Day | 285

ILLUSTRATIONS

Photographs by David and Lyn Hancock
Following page 106

David climbing to an eagle's nest in a huge cedar

*Lyn proudly holding an adult bald eagle she had just captured
as it was swimming across the channel*

With practice, it's possible to control an eagle with one hand

Proud and defiant bald eagle proclaims his superiority

David bands an adult bald eagle

The eagle's nest the easy way, by helicopter

Toweled immature eagles await banding at study base camp

California sea lions sunning themselves

A mother Steller's sea lion taking her pup to a safe place

*"Will I be oiled too?" asks one of the lucky, unoiled, sea lion
pups on San Miguel Island*

*David examines a sea lion pup on San Miguel that died in the
oil spill*

ix]

LIST OF ILLUSTRATIONS

David overlooking bird bazaars of Triangle Island

David being evicted from the back of a 6,000-pound bull elephant seal

David recording the strange sounds of elephant seals; many of the animals are molting

Murre colony

David with a murre chick he hatched

Murres, close up

Salal Joe's cabin and our plane in Barkley Sound

Our good friend Salal Joe hamming it up as he patrols the heavily forested islands of Barkley Sound

One of a pod of killer whales treading water in order to examine the Hancocks' skiff more closely

Islands of Barkley Sound

Our cozy retreat at Dutch Harbor, Barkley Sound

Following page 202

Sam investigating the bald eaglets

Sam in jail in Independence, Oregon (John Ericksen)

Sam: supercilious, alert, mischievous

Sam preventing Haida from leaving the yard

x]

LIST OF ILLUSTRATIONS

Sam cuddling up with his cat on the stove door

Peregrine falcon at nest

Four-week-old peregrine falcon from research breeding project being weighed at school

David uses a young peregrine falcon to illustrate a conservation lesson

Sam looking on greedily as Lyn feeds the eaglets

David and Lyn band a bald eagle

A bald eagle in flight

Measuring an adult bald eagle

The catch: an immature bald eagle flies off with supper in his talons

Views of bald eagles in flight

Maps

1 · Pacific coastline and the Channel Islands / *xv*

2 · Vancouver Island area / *xvii*

3 · Coastline from Vancouver Island to Queen Charlotte Islands / *xix*

ACKNOWLEDGMENTS

DAVID AND I WISH TO THANK ALL THE PEOPLE WHO HAVE assisted us so tremendously during our eight years of adventures studying the wildlife of the Pacific Northwest coast—the fishermen, loggers, mounties, lighthouse keepers, and coastguardsmen; those I have mentioned by name in this book and those I have not.

In particular we deeply appreciate the support given to our expeditions by the Canadian and National Audubon Societies, the New York Zoological Society, the American Museum of Natural History, the Canada Federal Fisheries Department, the U.S. National Parks Branch, and the B.C. Fish and Wildlife Branch.

We are also indebted to Avon rubber boats and Mercury motors for keeping us afloat when traveling by water; Mustang and Jones Tent and Awnings for keeping us dry on land; and Qantas, C.P. Air, and Pacific Western for keeping us airborne when we left the water.

Many thanks to our friend Edward Ward-Harris for his excellent advice and kind assistance; to David's parents and Anne Sinclair, who took over all my responsibilities and tied me to the typewriter; to Jill, Paulette, and Sue, who helped me with typing, mostly after midnight.

And of course most important, the one who started it all, my husband, David, without whom this book could never have been written; the one who invited me to share his life of adventure in what he calls God's Country; the one who introduced me to not only the puffins and eagles, but also the whales, otters, seals, bears, cougars, and all the other wonderful creatures who live there; the one who told me everything, then insisted I write it all down.

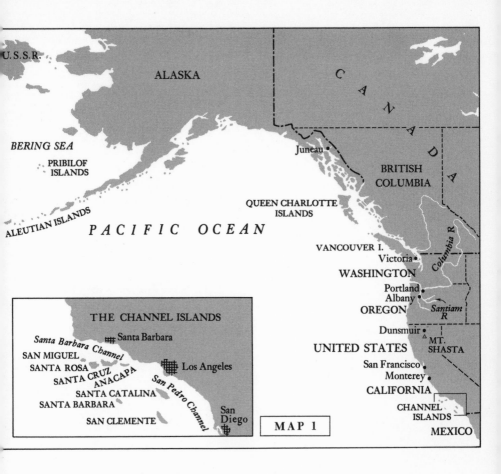

U.S.S.R.

ALASKA

C A N A D A

BERING SEA

PRIBILOF
ISLANDS

Juneau •

BRITISH
COLUMBIA

QUEEN CHARLOTTE
ISLANDS

ALEUTIAN ISLANDS

P A C I F I C   O C E A N

VANCOUVER I.

Victoria •

WASHINGTON

Columbia R.

Portland •
Albany •

OREGON

Santiam
R

THE CHANNEL ISLANDS

Santa Barbara Channel

Santa Barbara

Dunsmuir •

△ MT.
SHASTA

UNITED STATES

SAN MIGUEL

SANTA ROSA

SANTA CRUZ

ANACAPA

Los Angeles

San Francisco •

Monterey •

SANTA CATALINA

San Pedro Channel

CALIFORNIA

SANTA BARBARA

SAN CLEMENTE

San
Diego

CHANNEL
ISLANDS

MAP 1

MEXICO

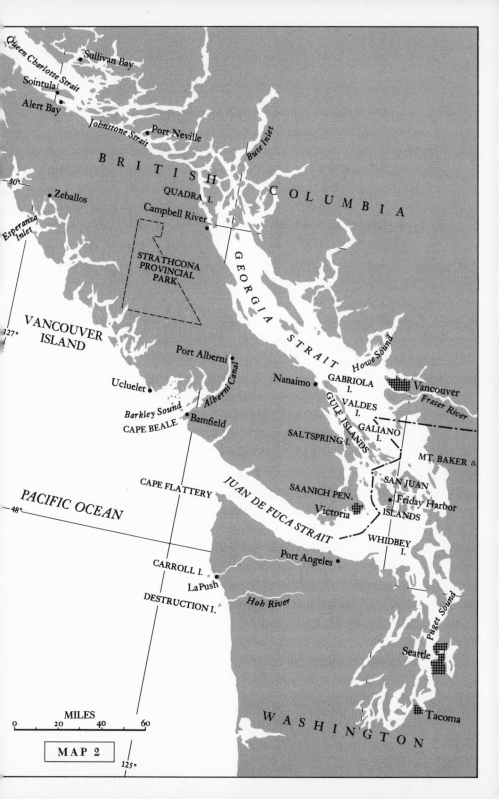

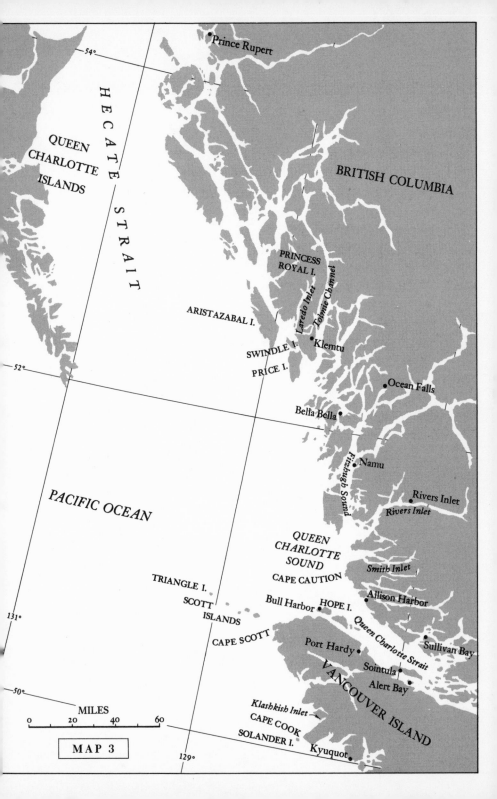

There's a Seal

in My Sleeping Bag

I

A Date to Count Eagles

EVERY TIME I STUMBLE BACK INTO CIVILIZATION, DIRTY AND drained of energy after some wilderness expedition with David, I vow that if I ever have the chance again I'll marry a sedentary nine-to-five office worker. But it takes only a hot bath and a good night's sleep on a mattress to make me realize how lucky I was to find, if not my ideal way of life, my ideal man—though I hasten to add that I very much doubt he would receive every woman's nomination for that title.

David is a wildlife biologist and film maker who believes strongly in the urgent need to preserve wild things, whether they are endangered species such as the peregrine falcon and whooping crane, patches of clean blue sky, or pockets of sweet-smelling mountain meadow. I believe in togetherness as a basic ingredient for a successful marriage, so I try to work with him even if it means, on occasion, sitting still for nine hours in an eagle's nest.

I have always loved the outdoors, and I survived some pretty harebrained experiences over a packed twenty-five years of living, but I never expected a marriage that would involve

3]

dangling precariously over sheer cliffs on remote sea-bird colonies in the Pacific, hacking my way with a machete through almost impenetrable rain-forest jungles, landing a dinghy with no previous experience in a twenty-foot swell on a beach slimy with seaweed, mothering sick seals and orphaned cougars, and mending an eagle's broken wing.

Ironically for a woman now so totally involved with animals, before I met David I thought of zoos as kid stuff. Just two months before I met him in Vancouver, while I was on a visit to New York, a friend suggested a day at the Bronx Zoo.

"Are you crazy?" I told her. "I stopped going to zoos when I was six. Let's try the Empire State instead."

Three months after that trip to New York I was married to a zoologist. Soon after I was off on my first wildlife collecting expedition to a remote island to collect puffins for the Bronx Zoo!

After a four-year hitchhiking jaunt around the world, I wanted to arrive home in Perth, Western Australia, exactly at 5 P.M. on Christmas Eve. It was a homecoming date I had planned from the start of my travels; besides I was broke and had only myself to offer as a Christmas present. To meet my deadline I had to board ship in Vancouver.

A few days before the boat sailed, while I was working as a very inexperienced waitress in a coffeeshop, David walked in sporting a profuse and multicolored beard, and with him was a glamorous blonde who also spoke Australian.

I conjured up a romantic picture of how the vivacious Australian girl had interrupted her world travels to marry a Canadian. But the next night David came in alone for a cup of coffee, then casually asked, "What are you doing next weekend?"

"Well, it's my last one in Canada," I told him, "so I intend to make the most of it. I'm flying up the west coast to an island in Howe Sound in a float plane."

David had another suggestion: "How about going to a farm on Vancouver Island by ferry, then flying further up the west coast in my float plane to count eagles?"

"Sorry!" I answered very decidedly. "But I have just too many things to do before going home. Besides, you are married." It turned out he wasn't married and that bit about counting eagles sounded so ridiculous that I canceled my plan in favor of his. I didn't really believe he had a farm or a plane or that he flew around in it to count eagles, but it did sound intriguing for my last weekend in Canada.

I had expected to spend the ferry trip through the breath-taking Gulf Islands ensconced in a cozy chair in the warm lounge. Instead I stood on a freezing foredeck, clinging to the railing as a bulwark against the strong winds, while David, oblivious of comfort—mine as well as his own—gleefully counted eagles and made notations in a little red book. The ferry route through Active Pass, he told me, was an important part of one of his eagle study areas, which covered the entire Gulf Islands. This area over which he flew every two weeks contained ninety-seven active eagle nests.

His parents on their farm near Victoria welcomed us warmly. His father, young-looking in his early fifties, was recovering slowly from a highway accident which had rooted him to the house for more than a year. His cheerful mother gave an impression of great strength. She seemed capable of anything, doing a man's work on the farm during the day and typing for David at night. She knew the meaning of hard work. As a girl during the depression years she milked a dozen cows before school. Now, as well as raising berries and flowers on their own farm, she packed eggs at a nearby poultry farm. It was from his mother that David developed his empathy with animals. When he was a child, a procession of pets led easily to an interest in wild animals.

As my surprises increased, my defenses crumbled. Not only did he have a farm and a delightful family but a light plane, a Piper Pacer on floats, his little GVE, which he used to tabulate the eagles in the Gulf Islands, the west coast of Vancouver Island, and Barkley Sound. Based at the University of British Columbia, he was studying the decline of an endangered species.

But, if I was lulled by the casual friendliness of David's home, I was due for a rapid change of mood as we took off on the plane trip from Elk Lake around the Gulf Islands. It was a nightmare. We flew just high enough to clear the treetops, so that David could see into the eagles' nests. At so low an elevation the strong winds severely buffeted the aircraft. Captive of the wind, the plane rolled, yawed, and pitched. My stomach never quite caught up as the plane dropped altitude with every downdraft. One moment the plane stood on one wing tip, the next moment on the other. Rain and sleet pelting against the fabric of the plane lessened visibility. It didn't make me feel much better when David commented that never had he been up in worse weather conditions.

Skimming over the tops of the thickly massed trees, David would swoop in on the massive stick nests that crowned the tall Douglas firs. The bulky nests, four to eight feet wide and two to ten feet deep, were the annual accumulation of branches, sticks, and twigs. Bleached white by the wind and rain, they stood out conspicuously against the dark canopy of the evergreen forest. Even more conspicuous were the white heads and tails of their adult eagle guardians. Then he would turn around, circle, and zoom back again for a second look. How he remained in control of so many things at the same time I found quite remarkable—the circumnavigation of every islet in that maze of islets, making detailed observations on tree, bird, and nest, even checking the accuracy of my minute pinpoints on his maps. Several times he glanced at one of my marks, took the pencil from my hand, and still making aerial maneuvers, altered my dotted position by a microscopic degree.

The constant change of focus between the horizon and the map on my knee, as well as the turbulent air conditions, was doing dreadful things to my queasy stomach. And then at a particularly bumpy moment I couldn't believe my ears: over the roar of the engine, David was saying, "How would you like to be Mrs. Hancock? You could cancel that boat back to Australia on Monday and marry me instead."

Later David told me he'd decided he wanted to marry me when he first saw my inexperience and inefficiency in that coffeeshop. When I survived the plane test, that clinched the matter.

I brought out every conceivable obstacle. His final exams were in a few weeks. And, anyway, I couldn't get married in Canada—my parents were expecting me. Oh, yes—and I was committed to a five-year contract to lecture in speech and drama at a teachers' college in Australia. And I couldn't cook. And I hated housework. And besides, I'd always believed marriage was something that happened when one ran out of exciting things to do. I wasn't ready to settle down....

But David had an answer for everything, and most uncharacteristically and to my own utter amazement I found myself saying "Yes." Then, as an important afterthought: "But you'd better stick with eagles, and not study fish. I can't even stand to look at them."

"You'll get used to it," he said grinning, and he banked sharply for a return to Victoria.

He wouldn't believe I meant to marry him, he said, until I myself informed his parents. "Why me?" I protested weakly. But a little later there I stood, rainsoaked and disheveled on the doorstep, and as Mrs. Hancock opened the door I blurted out, "How would you like this decrepit-looking creature as a daughter-in-law?"

David's mother smiled a wide, beaming smile. "Well, now, isn't that nice! So you flew off and got married?"

"No," David told her. "We're going to fly to Australia and get married there. Besides, there are some animals down there I've always wanted to see."

We hurriedly downed a glass of wine to celebrate before taking the next ferry back to Vancouver.

Hectic is hardly the word to describe the events of the following week. Getting married was to be only a brief diversion in David's life. First to the shipping office to cancel my passage on the ship's very day of departure, then to the bank to enlist the aid of a sympathetic and romantically inclined bank manager. Next to the university, to discover that all David's professors but one were away. The comment of that one astonished me. He threw his hands into the air, let out a whoop, and said this was the most refreshing thing he had heard since he arrived. David interpreted this as approval for a three-month postponement of studies. And lastly a flurry to scribble confusing notes on the back of two hundred already sealed overseas Christmas cards, outlining my sudden change of plans.

Somewhere around there I sent off a cable to my parents. "Met the ideal man. His name is David Hancock. Please prepare a Christmas wedding. Flying home. Love, Lyn." Never in all my letters home had there been the slightest reference to a David Hancock. Not surprising, when I'd only met him myself a few days earlier. But gallantly they replied with a cable of their own: "Wonderful news. Love to both. Dad and Mum."

After some hurried repacking we dashed to the airport to buy tickets—and came to a sudden full stop, as David realized that his passport had expired. Normally, arranging a passport through Ottawa takes an interminable time. But now even Ottawa moved fast. Unfortunately, though, in an attempt to make a good first impression on his prospective in-laws, Dave had decided to shave off his beard. I couldn't help wincing when I saw the result for the first time. I'd spent a few days getting used to one face and now I was presented with another. David's new face caused a delay in the passport proceedings: his picture didn't match up with his face sans beard. But finally all was clear, and at the airport, as we boarded the Qantas jet for Australia, I took off my furs and gave them to a friend. Next day we were to land in Perth in a haze of flies and blazing heat.

I tried to ensnare David into a round of dinner parties and wedding preparations but he would not be budged. Instead, he spent his time birdwatching with new-found (and bearded) naturalist friends, riding horses in the Darling Range, sitting in blinds at the top of gum trees photographing hawks and eagles, or meeting other zoologists.

Although charmed by her future son-in-law, Mum was horrified one day to open her refrigerator and find the freezer section crammed full of dead parrots. She was being initiated into a completely different way of life. She'd often harped on the theme "When will you ever meet Mr. Right, settle down and lead a *normal* life? One of these days it will be too late and you will finish up a lonely old lady with a parrot!"

Well, I finished up with Mr. Right, but with a wedge-tailed eagle. On our wedding day we received at the church one of David's most prized wedding gifts, a live, fully grown wedge-tailed eagle wrapped in a bright red bath towel, our gift from a veterinarian in Fremantle. The minister commented during the service that he was mildly surprised that the groom had been on time: he thought he might have been waylaid by the pigeons in the church loft.

Trailing tin cans and "Vive la Vancouver" signs, we rattled off in an old car to Waikiki Beach, south of Fremantle, and the neighboring islands to start our honeymoon.

Mine was not your average everyday honeymoon, but to my surprise I even enjoyed it. Certainly I found myself doing things I had never done before.

At low tide one day we walked across to Penguin Island, home of the diminutive fairy penguin. Penguins effectively illustrate a biological generality called Bergmann's Rule, David explained, which states that of any related group of warm-blooded animals the larger species are found in colder latitudes, and the smaller species in the warmer latitudes. Smaller bodies have more surface area in relation to their bulk, and therefore radiate heat faster; since they get cold more quickly, they must be confined to warmer climates. The emperor and king penguins, the largest species, are found in the frigid waters of Antarctica. The Galápagos and fairy penguins, the smallest species, are found in the warm tropical waters such as Australia.

We found these dapper little birds on leaf and seaweed

nests, in shallow excavated holes under camouflaged spiny brush. On one nest we found a fairy penguin still guarding her two soiled eggs. This coloration suggested they were near hatching, for initially the eggs are white, then slowly they become discolored with dirt. With disgruntled little barking noises, snaps from her formidable beak, and flogging flipper motions, she courageously withstood David's attempts to re-organize the eggs for a better photograph. There was no adult on the next nest, just a single three-quarter-grown young, already wearing the steely blue feathers that displace the gray down. Already he'd assumed the irascible haughty personality of his parent.

On Rottnest Island we fell in with a couple of Canadian zoology students who were studying quokkas, small nocturnal ratlike marsupials. Their objective was to trap the animals for banding and marking in order to learn the location of the animal's territory, its longevity, age, weight, and measurement.

Rottnest Island or Rat's Nest Island is so named because the early Dutch explorers of the West Australian coast thought the quokkas were rats. Rottnest Island was my family's favorite stamping ground, and I was pleased that it featured such a unique Australian animal. Before I met David I had always taken them for granted.

The first problem was how to capture a nocturnal animal that can get along better in darkness than its human pursuers. David sat in the front of the jeep, armed with a big dip net, and a miner's lamp with a spotlight strapped to his head. When the headlights illuminated a quokka on the road the animal was temporarily blinded and dazed. The first time, the quokka, as soon as it was accustomed to the lights, bounded off through four-foot-high Australian shrubbery. David in great enthu-siasm got out of the jeep and careered wildly into the bushes after it. The quokka turned to the right and David leapt over the thick spiny brush to cut it off. From the brush came shouts and yells and then silence. Eventually David limped back to the jeep with many prickles in his legs and rear end, but no quokka in his net.

Other quokkas were more cooperative and the hunt continued till eleven quokkas had been captured. At two in the morning we all retired to the cabin, where, so we were told, it was necessary to consume about four quarts of beer to replace the blood that had been lost in bounding across the dry shrub country.

While I slept, indefatigable David helped the other two Canadians to analyze, measure, weigh and band all eleven quokkas. Five hours later each animal had to be returned to the exact locality in which it was captured. This was most important because each quokka is very territorial and so might suffer reprisals if deposited in another quokka's territory.

On Dyer's Island we photographed and cavorted with one-ton sea lions. As we approached the sleeping colony of bulls David was careful to keep out of sight and very quiet. In fact he spoke rather harshly to me for not being as quiet as he thought was necessary. I was puzzled: I had often walked up to these sea lions on the beach without making them budge.

As I thought, David was used to the wild, easily scared sea lions of the North American Continent. On Dyer's Island the animals were completely lethargic and tame, a difference due to the fact that the Australian animals are on park land and completely protected from hunters, whereas on the North American coast sea lions are regularly attacked by people in passing boats.

On Garden Island we investigated the nesting sites of Caspian and bridal terns. Then, leaving the ocean, we climbed into the hills to the sheep country. David tried his hand at pit-lamping for rabbits and foxes, competitors and predators against the sheep farmer's livelihood. He was surprised to learn in Australia that pitlamping or spotlighting not only is legal but the accepted way of hunting. With the hot weather most animals only emerge after sunset to forage for food, so hunting with a light is essential.

Too soon we had to leave Western Australia on what I hoped would be a leisurely return to Canada. But the pace quickened. First stop was Adelaide. Here David almost missed his connection to Melbourne. I had no idea before that there were animals to study at the airports. While I stood minding the luggage David ran off to investigate the effects of bird life on aircraft.

The last call to board our Qantas jet sounded. No David. Should I go to Melbourne, and wait for him there? No, that would be disastrous: to lose one's husband when he was so new?

The boarding ramp was about to be wheeled away. I decided to walk to the plane; to stay on it or not could still be decided. With the help of a couple of volunteers to haul the luggage I reached the door to the plane. Still no David. I was on tenterhooks. The stewardess was insistent. The doors were about to close.

Then David came bounding across the tarmac, up the steps, and took his seat with as much aplomb as if he'd been the first passenger aboard. "I saw more bird species than I've ever seen before," he blandly remarked.

Around Melbourne and its environs David spent several more days revealing Australia's unique flora and fauna to me. I was ashamed of my ignorance: a Canadian showing an Australian the facts of Australian wildlife!

On Phillip Island we photographed koalas, those cuddly shy marsupials of the eucalyptus trees. In beautiful ferny Sherbrooke Forest we tracked down the lyrebird and David showed me how to approach close to a bird without disturbing it. The secret is not to let the bird see you looking at it: as long as the bird doesn't see your eyes looking, it thinks it has not been detected.

At Healesville we photographed other fascinating Australian animals: the unique and primitive egg-laying platypus; the brush turkey, which scrapes together great mounds of decaying vegetation in which it lays eggs that are incubated solely by the heat of decomposition; the beautiful rainbow bird and the flightless emu.

Poor David! Not even the animals of Sydney's beautifully situated Taronga Park could stop him from flaking out on the grass from heat exhaustion, and I administered a steady stream of soft drinks and ice packs. He recovered the next day, stimulated by the thought of searching for the white-breasted sea eagle, a cousin to the bald eagle he was studying in North America.

The French had just been offended by some aspect of Australian foreign policy, so flights to Tahiti were prohibited. For our return route to British Columbia we decided on Waikiki in Hawaii, where I looked forward to sun, surf, and siesta on the beach for the final week of our honeymoon, before resuming work in Vancouver.

I should have known better. For David, one hour's respite was enough. Other tourists in Hawaii strolled the beaches or sipped cocktails at hotels: the Hancocks looked for birds. In no time at all David had contacted some bird watching associates and we found ourselves looking through binoculars for the apapane, elepaio, iwo, and rice birds of the Central Hills of

Oahu or the gannet and booby colonies of the northwest coast.

Honeymoon over, I arrived back in Vancouver exhausted and dazed, to start married life with a man I'd known exactly one month. Considering the circumstances, I had to forgive him for forgetting my birthday in Sydney. I don't think I knew his either.

2

A Seal on My Honeymoon

"Congratulations, Lyn! You've just become the mother of a seal." David's words introduced me to the first of a long line of animals I was to meet in my new life. David had already raised several hair or harbor seals and had always been hoping for a chance to do a closeup study of a fur seal. Now within a month of our honeymoon I was to be initiated into living with animals.

We caught the next ferry to Victoria on Vancouver Island and within a few hours found ourselves in possession of a very emaciated and unhappy fur seal. With those shiny golf-ball deep-velvety eyes, Sam was impossible to resist. And since fur seals rarely survive in captivity, to keep him alive was my first challenge.

The northern fur seal is not really a seal at all but a member of the sea lion family. Sea lions are characterized by small external ears, long flippers, and great maneuverability on land. Seals on the other hand have no external ears, have short stubby flippers, and wriggle along on their tummies, wormstyle.

Fur seals breed on the barren rocky islands of the Bering Sea in Alaska. A million and a half or eighty percent of the existing population breed on the Pribilof Islands, which were discovered in 1786 by the Russian navigator, Gerasim Pribylof. Twice during the Russian administration the herd on the Pribilof Islands was threatened by annihilation: first, through failure to restrict the numbers of seals killed, and later by failure to protect the females adequately. The Russians are said to have taken more than two and a half million pelts between the time of the discovery of the islands and the sale of Alaska to the United States in 1867.

Fur seals were vulnerable to capture at sea as well as on land. Between the 1800's and 1912 huge fleets went out from Canada, the United States, Japan, and Russia to kill the animals on their southward and northward migrations. But the indiscriminate harvest was taking its toll. From several million, the population decreased to less than a quarter of a million.

In 1912, in what is probably one of mankind's best-honored treaties, the fur-sealing nations agreed to end pelagic sealing (killing them on the open sea), to cooperate in scientific investigations for management of the herds, and to share the harvest profits.

The success of this cooperative venture is seen in the fact that it is now desirable to decrease the numbers of breeding stock to reduce their natural mortality. When the number of animals increases beyond a certain level, the transmission of disease and parasites by close and constant contact increases out of all proportion. Furthermore, the greater number of females means greater competition for the available food, hence the increased distances that they must travel to find food to feed themselves and their young. There comes a point beyond which the female search for food takes longer than the pup can go without food. Thus the pups slowly weaken and starve to death.

Good management, therefore, means regulating the size of the herd for a maximum yearly harvest, yet with a minimum of losses from natural deaths. It is believed that half a million pups annually would be optimal.

With Sam as the stimulus, I became fascinated by the life history of the fur seal. In late April, after a winter spent in the open ocean, principally in the Gulf of Alaska, generally thirty to ninety miles from shore, the adult bulls start to haul out on the islands and select their breeding territory. The massive, arrogant-looking four- to six-hundred-pound bulls must content themselves with challenging other bulls, being challenged, and waiting, because the demure ninety-pound females don't arrive from their more southerly wintering grounds off southern California and Japan until early June.

Each bull may have between one and a hundred mates in his harem: forty is the average. The female chooses a desirable site for having her pup and then becomes part of the harem of whichever bull has taken possession of that area. The harem master is too busy defending his wives and territory against encroaching bulls or flirtatious bachelors to seek more females himself, and so the size of his harem largely depends upon the desirability of his territory to the females.

The female, impregnated the previous year, gives birth to her single pup within a day of her arrival at the breeding grounds. During the next week or so, the nine-pound youngster nurses on the mother's high-fat-content milk. Mother is mated for the following year, and then she heads out to sea to find food. For one to two weeks the young are left alone to sleep, fatten up on the tremendously rich milk they have stored in their stomachs, and try to avoid being crushed to death by a passing bull.

For over three months this sequence of feast and famine continues. By December the pup weighs thirty pounds and has learned to swim well on his own. Nursing abruptly ends and mother does not return. The pup must learn to find food himself. This means that he must first get to an area where there are fish and then must learn to catch them.

Unattended by mothers, half a million pups radiate out into the Pacific Ocean as far south as the Channel Islands off Santa Barbara, California, in search of fish and squid. Life does not come cheaply. About 15 percent of the pups produced each year are dead before the migration begins. Hookworm, infection, starvation, and injuries take heavy tolls. Up to eighty-five out of every hundred pups die a natural death before they are three years old.

Our Sam was destined to be one of these eighty-five, had it not been for our intervention. In some years dead or dying fur seals come ashore on the northwest coast in great numbers. In 1960 the United States Fish and Wildlife Service estimated that seven hundred carcasses were washed ashore during January and February in Washington and Oregon alone. Fewer seals come ashore along the British Columbia coast. It is suspected that during excessively cold stormy weather, storm-driven waifs have been unable to catch sufficient food to maintain their body heat. Some of the emaciated carcasses weighed only eleven pounds—a reduction to near birth weight.

Sam was almost dead, and too lethargic to eat on his own. Feeding became a fight. But, before we left Victoria to return home to Vancouver, David's mother, who had lived through several of David's seals, initiated me into the mysteries of force-feeding seals with a mixture of liquefied fish and antibiotics pureed in a blender—one of my most useful wedding presents, as it turned out. Three times a day for a month, David, protected by two coats and two pairs of gloves, wedged the snarling, squirming furry beast between his legs. While he kept the seal's jaws apart I squirted a plastic bag of the fishy, viscous mess through a thin rubber hose and down into his stomach. This method of feeding is called intubation.

Many a time the contents of that plastic bag squirted all over me before I perfected the technique. My mother had always despaired of my ever becoming the perfect housewife, but by wielding mops and brooms after Sam three times a day I soon acquired proficiency. Neither had I credited myself with much of a sense of smell, but oolachan, smelt, and herring did their best to stimulate one in me.

After the first week our patient was snarling encouragingly and the hollows between his ribs began to fill out. But as his health improved his disposition deteriorated.

At that time I was working as a substitute teacher in various schools and we were living in a four-room basement apartment. Every day and night David had to carry me out and in through the cement stairwell, our sole exit, for fear of Sam's shredding my legs. Sensing that Dave was unafraid, Sam made no attempt to bite him. Fur seals' social organization is based on dominance; the biggest bully wins. And Sam was determined to be on top from the start.

But enough was enough. After six weeks David decided that he was no longer going to be dominated by a twenty-pound biting monster. He pushed his double-jacketed arm and double-gauntleted hand toward Sam. In his accustomed manner, Sam latched on to it and started his very effective head-twisting movement to work his teeth through the layers.

David shouted, "No," and followed it by a sudden blow

across Sam's neck. Again the gloved hand was presented and again Sam grabbed it hard. The shout and the clout followed. The third time Sam hesitated, then latched on with renewed vigor. But David's third clout and the command "No" did the trick. The dominance order was successfully reestablished. Never since then has Sam seriously and savagely tried to bite anybody.

Force-feeding Sam with the plastic bag was simple compared to feeding him whole fish.

All my life I had detested the sight or feel of fish. I prayed that I'd never have to face the challenge of handling a herring. I thought I'd stipulated that David study eagles not fish before I accepted his proposal. He had refrained from telling me that everything he wanted to study, including eagles, lives on fish.

I remember vividly the trauma of the day I arrived home early from school, warded off Sam with my large handbag, and then bounced into the kitchen to be confronted with a roomful of fish. Herring covered everything—sink, table, floor. Scales clung tenaciously to wall, knives, and blender. Whole fish, strips of fish, heads of fish lay all over my kitchen in a malodorous array.

Picturing a whole army of fish on the attack I screamed and ran down the street. Although David later swore he had been called away in a hurry before he could clean up, I swear he tried to conquer my strong aversion to fish with what psychologists call shock tactics.

For a couple of weeks I conned various teachers into giving me a lift home. At the door I'd casually say, "You wouldn't like to come in for a cup of coffee, would you? And, er, perhaps help me shove a herring or two down my sea lion's gullet? You see, I have this aversion to fish and er . . ."

I knew I had to do something soon to cure my antipathy

before I ran out of friends. So, while David was away at university, I would gingerly unroll a newspaper package full of herring and force myself to stare at those goggling, cold, fishy eyes. Next I'd stretch forward a hand to touch the fish, first with a knife and then with the tip of my finger. At each touch I'd suppress a scream and instinctively my hand would shrink back. Trying to clear all aversion from my mind I'd close my hands on the slippery scaled body and pick it up, reflecting meantime what women will endure for love of a man.

David packed Sam off to the local zoo to see if the sea lions there could teach him to gulp down whole fish. No such luck. He wouldn't eat and he wouldn't behave. He was returned to us in disgrace.

Then one memorable day three months after Sam came to live with us, David tossed him fifteen herring and Sam swallowed them in quick succession. However, it was six months before I could tolerate handling a raw herring myself.

Now the arrival of a ton of herring at the front door is of little concern, and I can touch, cut up, and put pills in literally tons of all kinds of fish. I hasten to add that they must first be dead. As yet I've never had to finger a quivering live specimen. I am a mistress of artifice in dodging the issue. If we are scooping up herring from a wharf and some miss the net, there is usually a fascinated passerby to lend me a hand in picking them up.

The head in particular sends shivers up my spine. If I can hack that off first I can handle the rest. When David jigs cod I deftly move up to the bow of the boat and pray that he won't catch one. One of these days I'll probably slip into the water as the boat fills with fish.

Once Sam started eating, it seemed he'd never stop. In the wild, fur seals feed on squid and small schooling fish such as anchovy, capelin, and herring. Salmon occurred only in 28 out of 1,829 stomachs examined in a government study. Sam's normal diet became herring, vitamins (B_1 and multipurpose), and salt pills. To add variety we'd occasionally invest in hundreds of pounds of cod, halibut, and sole. Herring is a fish that contains an enzyme called thiaminase which breaks down thiamine in the body. On a steady diet of herring Sam would lose thiamine which had to be given separately in the form of B_1 additives.

We tried various methods of keeping pace with Sam's appetite. We chartered fish boats, bought from commercial fishermen and supermarkets, even netted herring, smelt, and oolachan from the beach with purse seines and gill nets.

Ironically, all this fish had to be frozen first before Sam could eat it. Fresh fish offended his sensitive digestive system. And, since a regurgitating seal with diarrhea is to be avoided, we had fish tucked away in practically every neighbor's freezer, forty pounds of cod across the alley, two hundred pounds of herring on the left, a hundred pounds of smelt on the right. When I overheard two of the neighbors discussing selling their freezers, we decided to discontinue their voluntary services and try someone else. Sam's preferred food even now is herring—old, freezer-burned, dehydrated, and browned with age. Once they are thawed and watered he relishes them and grows fat.

We don't know what it is about fresh herring that causes stomach upset. One of the effects of long-stored and dehydrated herring is the change in the structure of the herring fats and oils. Quite possibly Sam simply gets adjusted to the aged herring which is our most available food supply and he is upset by the different oils in the fresh fish. Obviously there can't be anything too much wrong with fresh herring because wild fur seals would seldom find in the sea fish that was old, freezer-burned, dehydrated, and browned with age.

With neither of us earning a regular salary, finances were always a problem. As Sam approached forty pounds we found we were paying more for his food than for our own. Herring cost us as much as twenty-nine cents a pound and some days he was eating well over fifteen pounds. Sometimes we'd take Sam down to the beach where people were fishing for smelt, and let Sam and his appealing eyes earn his own handouts. Sometimes we tried fishing ourselves. But the amount of smelt that would send the average family home happy was just a mouthful for our Sam.

David's first attempt netted one-tenth of one meal, six smelt. But he did catch a bonus, one live cutthroat, which he put in the bathtub with Sam to see how the seal would react to live fish. At first sight of the trout Sam jumped right out of the water. At times like those I wondered how he could have managed to traverse the ocean even as far as British Columbia.

A problem with all seals and sea lions in captivity is blindness due to a gray film that forms over the cornea of the eye. Left long enough this film becomes permanent and irreversible and effectively prevents light from entering the eye. Subsequently the seal or sea lion grows blind.

David had found out with his first harbor seals that growth of the film started when the seal had been several weeks in fresh water. He discovered that the condition could be cured by one of two methods. Either the seal could be placed immediately in salt water or else taken from the water so that the only liquid in contact with the eye was the salty excretions of its own tear ducts. In both cases the lack of fresh-water culture media seems to destroy growth, which David believes is bacterial. He has made these suggestions to several zoos and they have been able to prevent blindness in their seals or sea lions.

The first two questions we encountered as soon as Sam's

stream of visitors began to arrive were: "Why isn't he wet and slimy?" and "Won't he die out of water?" Most people see seals in water so they find it amazing how easily he moves about on land. He is only wet if he has been swimming. He can go months without being in water (although water is desirable) and he is not at all slimy. Sam wears an extremely expensive fur coat and I doubt if any woman would admit that her fur coat was slimy. Only the outer guard hairs get wet and lie smooth and shiny. Underneath these is the tremendously thick layer of fine fur.

Often we smuggled Sam into the pool outside the Education Building at the university. With the typical agility of a sea lion he used his unusually long, glovelike flippers to propel himself through the water. He certainly enjoyed the Education pool. To the amusement of his spectators he swam, splashed, and scratched. Once out of the water he spent a lot of time rubbing and preening his luxurious fur coat with the three long claws on the upper side of his back flippers. So often did he do it that I was tempted to nickname him Itchy.

Yet, contrary to the belief of most observers, he doesn't have fleas and lice. His fur is too thick for them to live in, and fur seals are meticulously clean. The constant rubbing keeps their fur free of dirt. Indeed, it is a life-and-death matter. If a seal's fur or, for that matter, a seabird's feathers become dirty, they would lose their waterproofing and quickly die from loss of heat to the surrounding water.

We took Sam everywhere with us. He looked forward to visiting people who owned television sets. Dogs, which depend almost entirely on scent to recognize living creatures, will seldom give more than momentary attention to the flickering figures on a television screen, but seals and sea lions have excellent eyesight. Sam liked to move close to the screen to watch the action. But invariably he was startled and leapt backward when a closeup flashed on. Probably it was just the novelty of television that attracted Sam. It would wear off for seals just as it does for humans.

Sam loved going for rides in the car. His favorite position was lying on our shoulders or sitting on our knees with the front three-quarters of his body hanging out the window like a signal arm. Many were the double takes of other motorists waiting in the next lane at some stoplight when they glanced up to see a seal with flippers hanging down to the door handles and nose pointed up in the air.

Returning to the parking area from shopping one day we saw our car completely surrounded by people. We pushed closer but no Sam could be seen. The windows were all white. Peering through the partially lowered window, I gasped. Sam's face was a sight! White icing glistened stickily on each whisker. Two dozen freshly iced doughnuts had been chewed and tossed about the car in his favorite game of "Kill the fish." Gooey icing and cake had been thoroughly smeared over the windows as he sought for attention from his audience, and the seat, steering wheel, and dashboard remained sticky for weeks afterward.

Just as I was ready to blow my top, with a twinkle of his soft velvety eyes he flopped over to my window and planted his "sweet" whiskered face against mine. The onlookers couldn't contain themselves. And neither could we.

When we went visiting, Sam was usually left outside in the car, although it varied according to the tolerance of those we visited. Our socializing decreased when Sam learned how to blow the horn. Many a neighbor was rudely awakened from a midnight sleep by Sam playing his loud one-flipper tattoo. He was never still. Completely disregarding people or property he clambered from back seat to front seat, along the dashboard, over the steering wheel, then front seat to back seat again.

Forever curious, he was skilled at opening and closing doors. When left in bathrooms he invariably ended up inside the wastepaper bin or laundry basket.

We'll always remember the night we returned to find Sam had somehow gained entry to the house from the backyard. Although I felt some misgivings nothing could have prepared me for overturned chairs, tea towels pulled down from the rack, a sack of flour dumped out on the kitchen floor, and a set of white floury flipper prints leading to the living room, where drapes, cushions, and lamps were a jumbled mess in the middle of the carpet. With a vehement "Oh no!" I dashed to look in the bedroom to find Sam, the culprit, at the height of his havoc. Curtains and dresses had been pulled down. Sheets and blankets were ripped from a completely denuded bed. And there lay Sam, complacent and pious, waving a flipper from the middle of a warm wet mattress.

Reveling in his totally undeserved attention but with a slightly sheepish toss of his head, he lunged outside after the herring in David's persuasive hand. Unable to face extensive housecleaning at midnight, I zipped our sleeping bags together and laid them out on the back lawn. We would share the fresh air with Sam.

I went back inside for a few minutes. When we came out to the lawn again I realized Sam would continue to attract our attention for the rest of the night. A shaft of moonlight glistened on a pert, saucy nose, black shiny eyes, and long stiff whiskers as a mischievous seal poked his head from the recesses of the sleeping bag.

He loved pushing boxes around the house. This was acceptable except when it was the garbage box. If the box was large enough he sometimes buried himself inside it. Guests met at the door by a box apparently moving around the house by itself would invariably be taken aback, to say the least.

To seal-proof the house would have meant removing the furniture: Sam would climb everywhere—up on the table, over the typewriter, on top of the beds. He'd sweep off the toaster, knock over the clock, pull up the venetian blinds, and peek out the window at the neighbors.

The complete extrovert, Sam wanted constant attention. Since most of my time in the house was spent sitting at a typewriter, this meant Sam climbing up the back of my chair, hugging me with his flippers, tickling me with his whiskers, and then scrambling over my shoulders. Typing was definitely interrupted. If he considered himself neglected he'd tear up the eiderdown or rip up paper bags to capture attention.

His favorite game was Seals and Matadors. Sam really enjoyed it when I got down to his level and frolicked on the floor. I pretended I was a seal and we chased each other all over the house. I was no match for his agility: he could almost turn somersaults with his streamlined body and nimble flippers. But I could usually fool him by jumping over his head and carrying on the chase in the opposite direction.

The game would end when Sam was too hot. He'd then seek out the cool wood that bordered the rug and spread out his four long "black kid gloves" in the air like fans.

With David busy at the university and me teaching at school we obviously needed some playmates to divert Sam. Luckily at this time we met Hilary Stewart. Her sister-in-law Anne taught in the next room to mine at Maple Grove Elementary. When Anne first saw Sam she immediately commented, "You must meet Hilary. She'd adore Sam."

Hilary is a very special kind of girl, a treasured friend over the last seven years. She is the kind of girl who thinks nothing of picking up a parcel for you at the airport to find out later that it contains a pair of lynx, which she obligingly loads into her car and delivers to the door; who takes a cougar home to her apartment for a few weeks until her landlord gives her notice; who relishes building a fence, laying cement, or just "fixing" things. Over the years Hilary has never been fazed by any of the creatures we brought to her door, whether they were sea lions, cougars, raccoons, apes, black bears, or coatimundis. Not that her apartment lacks wildlife of her own: she specializes in salamanders, lizards, frogs, and snakes when she isn't boarding the Hancock menagerie. Although she worked as the art director of a local television studio, she likes to forget it most of the time and looks forward to her life outside it.

Hilary's apartment is typically Hilary. Her paintings hang on the walls; a collage of shells and driftwood stands beside the door. A chain of rusty iron dangles above a skull in the fireplace. Eskimo and Indian artifacts crowd the shelves, jostling for space with books on hiking, survival, archeology, plants, and animals.

When Sam first flippered through her door I knew we'd found a handy seal-sitting service. She frolicked on the floor with him in gay abandon while I kept a nervous eye on her "things."

The canvasser for the United Appeal was not as reckless when Sam met her at our door. The Red Cross described the meeting in their *Newsletter:*

EVEN SAMMY HELPS UNITED APPEAL

As many of us are well aware, canvassing for the United Appeal can be an interesting and rewarding experience, providing the opportunity for meeting many new faces.

Residential canvasser Mrs. E. Jackson met a new face just the other day. It belonged to Sammy, a seal, who greeted Mrs. Jackson when she was making one of her calls. Our canvasser, somewhat taken aback, retreated down the steps and negotiated with the seal and his owner from the front lawn.

The story has a happy ending since Mr. Hancock contributed to the United Appeal and Sammy gave Mrs. Jackson an affable kiss—surely a genuine Seal of Approval for United Appeal.

Although I am not enamored of teaching tricks to animals as if they were circus performers, a few commands were necessary for Sam to know. David easily taught him by the reward system to Come Here, Go Over There, Go a Little Further, Stand Up, and most important of all, to respect the command No. From Sam's point of view the house was a world of no-no's. Sam was intelligent enough to realize when he had done something we disapproved of, but also intelligent enough to continue to do it when we were not looking.

The greatest no-no concerned toilet training. Seals in the wild perform such functions without thought or premeditation, so it was difficult to make Sam feel "guilty" about "going" in the wrong place. If the door was left wide open, he went outside. If it wasn't he definitely looked agitated, but being a species of sea lion that doesn't often vocalize we couldn't teach him to "bark" at the door. And because Sam often got nervous in strange situations it was difficult to know if he had to "go to the bathroom" or merely was unsure of his surroundings.

Such a situation took place at the Bayshore Inn, one of Vancouver's most modern and fashionable hotels. Sam and I share the doubtful honor of being booted out of it in a hurry. When I first arrived in Vancouver to look for a job the Bayshore hired and fired me all on the same day. Sam was the next Hancock to grace its sophisticated corridors. A well-known construction company wanted Sam to attract attention to their Seal Joints at a convention they were holding at the hotel, and David packed Sam off to the Bayshore to earn his next month's meals.

Sam attracted attention all right, but probably not the kind that was wanted. Something scared him and off he galloped to the barber shop, the foyer, the dining room, with David giving chase. Sam entered one door of the beauty parlor and two screaming ladies with curlers falling from their hair fled from the other. Three times Sam left the convention to flipper through the hotel, to the delight of the guests but the frustration of David, the hotel management, and the company that had hired Sam's services.

Eventually the manager cornered David and informed him that Sam was contravening the health regulations by being in a public place. Sam was led off to the exit, but still nervous and agitated he managed to get away again. This time he led everybody a merry chase through the corridors to the main entrance of the hotel, where he deposited a pile on the Bayshore's best carpet. In times like these you need a healthy stock of paper towels and a copious supply of soothing talk.

I have to admit that we weren't really doing a very good job in training Sam to do as he was told. That is the problem when you have a pet and not a circus performer.

One morning Hilary phoned in desperation. "Lyn, can you bring Sam to the studio tomorrow morning for a children's show? Some arrangements have been canceled and we need something exciting in a hurry."

It had been David's policy to keep Sam away from T.V. studios, but I made a quick decision to take Sam myself. This meant getting permission from my friendly principal for a brief absence, washing and combing Sam in the bathtub, lining the car with papers—and, first of all, staying up all night worrying about it. No sooner had I got started the next morning than the car broke down on the highway. Should Sam and I hitchhike together, or should I go out alone and then tell the driver I had a friend? Before I had time to contemplate the alternatives or the car's troubles a kind woman came to the car door to offer a lift. She was enchanted to help a sea lion in distress.

At the television studio I was to be interviewed in a school-room setting with six-year-old children who were to look delighted to have a seal at school. Unpredictably, Sam was a great hit with the parents but not with the children. He persisted in going after their legs under the desks while I crawled along the floor pulling him back by the flippers because the tiny tots weren't looking sufficiently appreciative of his attentions. The interviewer crawled after me asking questions, the producer crawled after Sam to herd him into camera range, and the cameraman crawled after all of us. Meantime Sam refused to do a thing he was told. He wouldn't wave, sit, or stand; he wouldn't even eat a herring! Everything was completely unrehearsed and the audience loved it.

To the oft-asked question "Is Sam housebroken?" we would reply, "He certainly is. He's broken four houses." We moved from basement to attic to duplex then finally to a bungalow with a large backyard—four moves in the first four months of our marriage.

At one stage in our search for accommodation when the honest approach, "Of course, you wouldn't mind if we kept our fur seal in the back . . .?" wasn't producing results and we had a six-day deadline to get a place to stay, we changed our tactics.

The landlady's interview went smoothly and nobody had mentioned pets. The deposit was given, and then just as we were leaving she called out, "Oh, by the way, I don't allow cats, dogs, or children. You don't have any, do you?"

David and I exchanged glances. "No, we don't have any," David replied in an offhand way. And, clutching the receipt for our deposit, we walked away. Perhaps we might get in two months' occupancy.

No such luck! Unfortunately we took possession of the duplex a night early, with the previous tenant's permission. That night the landlady came around.

"Are you the Hancocks with the pet seal? I won't have a seal on my property. No animals, no children, no visitors with animals or children" was her adamant verdict.

And since this was one day before the beginning of the month, she gave us notice there and then. She couldn't have been expected to know it, but the seal in question was at that moment in a trunk she had chosen to sit on during her lecture.

Yes, a seal in the family can be a problem.

Then out of the blue we came upon a landlord who had a pet Siamese cat named Sammy. Sammy-the-Siamese's owner, a veterinary supply dealer, was delighted when a zoologist with a pet seal wanted to rent his cottage. I'm not sure he was all that delighted later when four cougars, a dog, and a couple of lynx moved in—but then that is another story.

3

Nine Hours in an Eagle's Nest

DURING OUR FIRST WINTER IN VANCOUVER DAVID CONTINUED the regular census flights around his three study areas, the Gulf Islands, Barkley Sound, and the west coast of Vancouver Island. Meanwhile I seal-sat, taught school, and handled the correspondence of a small army of volunteer eagle observers from many walks of life throughout the province. David's work at U.B.C. allowed him to do the two things he liked best—piloting an airplane and studying the bird which had brought us together, the bald eagle.

His interest in eagles had begun during his high-school days when he became interested in falconry. He had trained several smaller hawks but always marveled at the lofty eagles, and it was the flight of eagles and hawks that inspired him to try flying himself. He qualified for his commercial pilot's license in Grade Twelve, and after several summers working in a local gravel pit he earned enough money to buy a small Taylor craft, then later a Piper Pacer on floats.

He debated whether to go to a university and become a wildlife biologist or whether to become a commercial pilot with nature study as a hobby. He chose the latter, but after three years of flying he enrolled at the university: the ability to fly gave a new dimension to a career in wildlife biology, and especially the study of his favorite bird. His flying experience proved that he could locate eagles and their nests from the air. By observing the habits of many eagles from the float plane he could study the habits of a whole population of eagles.

He began a study on the ecology of eagles, which included the seasonal population, movement patterns, and subsequent densities of wintering and breeding birds, productivity and nesting success, the feeding habits, and the effect of various habitat components on the distribution and abundance of the bird.

In other parts of North America biologists have faced a lack of birds to study. But in David's research area he was able to observe hundreds of eagles along thousands of miles of shoreline, from the state of Washington up the Columbian coast to the Alaskan border.

It was now spring. David and his assistant, Lionel Hughes-Games, were camped in Barkley Sound for the eagle breeding season, in order to continue detailed and sustained observations.

That first spring, while I waited for a permanent teaching position, I was able to join them. Sam stayed with David's parents in Victoria. David flew down to pick me up.

This was to be my first expedition into the wilds.

I laugh now when I remember the equipment I chose for that first trip—the great quantity of unsuitable food, the excess clothes, such luxuries as shampoo, night creams, combs, and mirrors. My plan was to make our cabin on a tiny island in Barkley Sound a real home away from home. And I, playing the part of the young bride to the hilt, was determined to make it so.

I finished packing the camping and food supplies into the plane as David filled up with gas and oil. On top of the gear I carefully placed four freshly baked desserts—a cheesecake, chocolate cake, lemon loaf, and date squares. The ropes were untied, David pulled the starter, and the float plane started to pull away from the dock. At the last instant I waved goodbye to a friend who was filming our takeoff and jumped from the dock to the aircraft. Once inside I got work maps ready while David checked the engine.

"All ready," I signaled. We started taxiing along the surface. The water was glassy calm and our speed increased. David eased back on the stick but the plane didn't lift into the air. Then he tried lifting one float out of the water at a time. Still the craft wouldn't break free from the smooth surface of water. David made two circles on the water and then cut across his waves. Still we scooted along the surface: we just didn't have enough lift. Back to the dock we taxied, a little embarrassed.

"That was a short trip. See many eagles?" our friend greeted us, a little annoyed because she had taken a whole roll of movies of our taxiing and now had no film left for the take-off.

There seemed only one thing to do. Reluctantly David unloaded the four cakes from the plane. I sacrificed some clothes, makeup, and a mirror. It was just enough.

Soon we were flying along the coastline skimming two hundred feet over the beach near the trees. We intended to survey a hundred miles of coastline from Victoria up the west coast of Vancouver Island to Barkley Sound. This coastline is relatively straight, more exposed to the Pacific and lacking the myriad islands of David's other two study areas. In eighty miles of shoreline, there were only a couple of lighthouses and one almost deserted Indian village, Clo-oose.

Seabirds were plentiful along the route: their different tactics of avoiding us were quite distinct. Gulls and scaup ducks scattered to each side of the approaching plane. The larger black cormorants and the small saucy buffleheads chose to dive into the quiet depths of the sea.

"One adult . . . two small downy young and adult by empty nest . . . immature flying . . . two adults above nest with one medium-sized young." As David called out his observations of eagles I carefully recorded them on the maps.

"Adult on nest containing—oops, the adult was standing in the way." David banked the aircraft tightly to the left for another look at the nest. This time he planned on putting the nest between the aircraft and the adult perched on the nest edge. David sideslipped the plane within fifty feet of the huge nest and the female eagle glared at us as we passed by.

"Two—no, *three* small downies," David yelled and I smiled back my delight. "I'd better check again. This is most unusual."

David banked the aircraft steeply. Sure enough, there they were—three small eaglets covered with downy gray feathers, and about ten days old. In over a thousand nests he had checked, David's previous flights had shown that about half the nests have only one young. Twenty-five percent have two young, and the remaining twenty-five percent produce no young.

It was at the entrance to Alberni Canal that I experienced my longest moments of sheer terror. As we rounded a point, without warning we flew into a living curtain of seabirds. The air was dense with thousands upon thousands of gulls, cormorants, eagles, and many others that my paralyzed brain couldn't identify. We had chanced upon a tremendous herring ball in the sea below us—a churning mass of small herring driven to the surface of the sea by predatory fish below, which had brought the entire bird population out in full force.

David tried to steer around the gyrating mass. Birds whizzed by on both sides of the plane, and above and below us. Miraculously, our small plane sheared through the turmoil of moving birds and we emerged unscathed on the other side. For minutes after the encounter I was still in a state of shock. If only one bird had hit our propeller, we would have joined the herring ball in the sea below. Calmly, David continued to count eagles.

We rounded Cape Beale and entered Barkley Sound. Before us lay an eagle watcher's delight, nearly two hundred islands and islets covering an area of about four hundred square miles, among which David had located 170 active nests. But, for me as a recorder, this maze of islands was more of a nightmare than the more open areas of the gulf, and certainly it was so for David as pilot and observer.

Somehow he had to circle each island to locate all the eagles, check the contents of each nest, miss the trees, then grab a glance at my map to see if I had marked the exact dot in the right quarter-inch space, then orient himself in the right direction and swoop on to the next nest.

Eagles were everywhere, and David hardly stopped shouting his observations. I had to keep looking out the window to keep track of where we were, then focus back on the map to record David's observations. For many of the male assistants who used to accompany David on these flights before our marriage the continuous change of focus had often caused motion sickness. David still teases me by telling me that the only reason he married me was that I didn't get airsick on that first flight.

"There's Lionel in the boat," David shouted. He pointed to a tiny rubber boat plowing through the waves toward the camp at Dutch Harbor, which was to be our summer headquarters.

Like many romantics I have always wanted to live on an island. Although we had only borrowed Dutch Harbor for the eagle-breeding season I found it a delight. Small enough to walk around in less than ten minutes, it contained a tiny two-roomed cabin with a wood stove and some sparse furniture. There was even a well and a rough toilet. The weather, warm and sunny, illuminated the colorful Indian paintbrush flowers that danced in front of the cabin in the light breeze. For the next two months cold sea fog and pouring rain alternately obliterated the landscape, but I will never forget the sunlit enchantment of that first day.

Our camp was at the mercy of wind and tide, with no place to tie up the float plane. As soon as we had landed in the sound and deposited our gear at Dutch Harbor David flew over to nearby Turtle Island, where his old friend Salal Joe lived on a float amid all his worldly beachcombed possessions. Here we could tie up the plane and we would commute from island to island in our leaky aluminum dinghy.

Commonly known as Salal Joe, Joe Wilkowski was the hermit of Barkley Sound. David knew him well. Joe had rejected the life of society and hoped to spend the rest of his life on his raft roped to Turtle Island. He earned a meager existence digging clams for the gourmet market or collecting the evergreen huckleberry and salal for florists in Victoria and Vancouver. Once a week he'd pack the brush into his flat-bottomed driftwood scow and go out to meet the *Lady Rose*, a small coastal freighter that plied between Port Alberni and Ucluelet at the northwestern entrance of the sound. Other times he would cross stormy, exposed Imperial Eagle Channel to deliver his clams to the fishing village of Bamfield, which guarded the southwestern entrance of the sound. This and the occasional trip into the nearest settlement of Ucluelet for gas, rum, and groceries were Joe's only human contacts.

In this era of people, pollution, and panic, his could be the ideal life—shelter and transportation fashioned from the beaches; food, crabs, mussels, oysters, and abalone, gathered from the well-stocked islands in the backyard; for entertainment, fishing, hunting, or watching the deer, the killer whales, and the bears that play on the doorstep. In Joe's words, "If I had a million dollars I'd buy lots and lots of grub and go on living right here. You can't have a better life."

Joe didn't quite fit into this picture of rustic perfection; he disliked such easily obtained delicacies as crabs, mussels, oysters, and abalone. He fished, but only for his half-wild cat, Chico, for whom he boiled cod or salmon, served with salt, pepper, and garlic. Joe himself preferred to eat fruits and vegetables that he grew himself beside an Indian midden on a nearby island. Every few days he made bread and buns which he called cement, but David, the world's greatest lover of baked goods, will affirm that Joe's bread is not related to cement in any way.

Whenever David tied up his plane at Salal Joe's he always slipped a bottle of rum down the hollow back of his wooden bench. Occasionally, when David and his assistants stayed overnight, Joe would keep them up all night listening to Radio Moscow or questioning the university students on dubious philosophical points until he'd drunk himself into oblivion. Having heard of his violent temper, reliance on rum, and dislike of women, I had some misgivings about meeting him, but no sooner had I stretched out one foot onto his float than he was there to hold my hand. As I stepped from the plane he removed his tattered woolen cap, bowed deeply, then caught my hand in both of his in a powerful double handshake. David was utterly flabbergasted.

I needn't have been so prejudiced by outward appearance and preconceived notions, for Joe was a true gentleman. He surprised me again the next day by coming over to Dutch Harbor with a loaf of bread, a comb, and his shaving mirror to replace my makeup bag which had been left behind on the dock in Victoria. Despite the fact that I didn't bother to use such luxuries for the rest of the summer, I was touched by Joe's thoughtfulness.

When I spent longer with him I found he was the most fastidious man I knew. He preferred to cook his own meals. He liked his fish to be browned on top and in one piece. He refused to cook potatoes or wash dishes in the easily obtained sea water available at his front door, preferring a long haul to get fresh water from a neighboring island. He cleaned his floors twice a day on his hands and his knees because "dust rises if you sweep." He wouldn't tolerate frying pans placed on the floor or cans thrown in the sea, and when we gave him a pet crow it had to be kept out of the cabin. Periodically Joe would scrub the logs of his float free of algae and sea anemones.

Salal Joe might be my friend but salal, a dense shoulder-high shrub, was an enemy I fought every day in the rain forests of Barkley Sound. The superabundant moisture, the almost incessant rain, the sea fog which rolled in every morning during the summer contributed to the development of huge long-lived trees like the towering Douglas firs, and the massive red cedars, and the shade-tolerant western hemlocks. At high tide the dense mass of foliage seemed to grow straight out of the water. The receding tides revealed a table of slightly shelving rock from which the forest grew. The treeline was razor sharp at water's edge. Once we had left our boat and penetrated into the eerie gloom of the forest interior, progress was almost impossible. Imagine a maze of tree trunks, draped with lush hanging gardens of ferns, mosses, and lichens, meeting a tangle of sturdy salal pushing up from the ground. Imagine clambering over rotting nurse logs, the fallen rotting trunks from which the new young seedlings grow, squishing down into the sodden spongy moss cushioning the ground. It would take more than fifteen minutes to cross the length of an average living room.

It was in such a jungle that David and Lionel had to work. Their first job was to build a blind, suitable for close observation and photography, at the top of a tree which looked down into a nest tree containing young eaglets.

Three blinds were painstakingly built in the pouring rain of my first British Columbian summer—a far cry from my months of lazy lounging on the beach in the baking heat of an Australian summer. First we scrounged planks and driftwood from the beaches. Then we blazed a trail uphill from the shore, and hauled the lumber up through the dense undergrowth to the base of the tree. Planks, saw, hammer, and nails then had to be hauled up more than a hundred feet to the top of the tree. And that meant not a simple haul straight up to the top, but threading all these things through and over and under the thicket of abundant vegetation.

Lionel finally punched spikes in the tree trunk so that I could join them for a few hours of eagle watching, sitting crouched and soaked at the top of the tree. Although Lionel assured me that the spikes made it so easy that even his grand-

mother could have managed the climb, I never willingly volunteered for blind duty.

Depending on the weather, Lionel or David climbed to the
top of a couple of trees each day to collect information on the
birds and the nest and to band the young.

To reach the top of one tree, I remember, took David eight
hours. To each leg he strapped a climbing iron, with long spurs
to dig into the thick bark. A long rope with a steel core for
strength was tied to his belt and looped around the tree. By
kicking the spurs into the bark he could "walk" up the tree,
gradually working the rope up the trunk. A logger would have
cut off each branch as he came to it, but David wanted to protect the tree, so he would balance on his spurs perhaps a hundred or a hundred and fifty feet up, unbuckle his belt, lift the
top over the branch then resnap the buckle into the belt. Sometimes he would use two belts, tying the second one above the
branch, then undoing the first one.

Once above my head, he was soon lost in the dense growth
surrounding the tree. The most difficult part was when he
reached the huge mass of two- to four-foot branches carefully
woven in place to form a bulky nest perhaps seven feet in diameter, large enough to hold several people. He had to climb
out around it and over into the nest. The center of the stick
structure holds soft grasses and moss to cushion the eggs. Year
after year the eagles return to renew the old nest. When one
examines closely the intricacy of the huge, ever-growing nests,
one thinks that the parent birds must spend a long time in construction. Yet two of David's eagle observers have reported
that they have seen eagles build the basic nest structure in only
two days.

Bald eagles mate for life, but if one mate dies, the survivor
will take a new partner.

Amazingly, the parent birds made no attempt to protect their young. As David approached the top they would circle the tree and chitter in alarm, then leave to perch on nearby trees until the intruders left. Grappling with a nearly full-grown eagle isn't easy. Balancing on the nest platform, David would grab its feet before it would grab him. Sometimes two eagles in the nest would defend themselves very adequately by flapping their huge wings. A blow on the side of the head could easily have toppled a man to the ground, a hundred and fifty feet below.

My job was to crawl on my stomach around the base of the tree and collect food items, such as feathers, bones, and rotting flesh, to identify what eagles eat. I would then clear a space around the trunk of the tree to make it easier to keep David's ropes untangled and to handle anything he might lower down to me. Sometimes he would lower a small eaglet for me to measure and band, while he photographed his observations on the size and composition of nest and tree. Luckily for me, as well as to protect the bird, the eagle would be wrapped in one of my bath towels for the downward journey. I would take the eagle out of the dangling banding bag, weigh it, measure its wing span, body length, foot width, and feather length and finally place an aluminum band around one leg, giving a unique number and the address of the United States Fish and Wildlife Service, in Washington D.C. For each band I would fill out a card with information about the bird, the date and place where it was banded, and the eagle's estimated age. If later the bird was captured or found dead, the band could be sent to Washington to give information on the bird's travels and its age.

When all observations were made I would wrap the eagle again in the bath towel and signal to David, who would haul it up to the nest again. Sitting in soaked jeans in the salal and fighting off the mosquitoes, I wrote long letters to my mother to describe my new life in Canada.

One day David brought a three-week-old eaglet back to the cabin. "This is a job for you, Lyn. You've mothered a sea lion, now have a go at an eagle. While Lionel and I are at the blind you can look after Little Jesus."

Contrary to popular opinion, an eagle, especially one only three weeks old, is really a delightful creature. Clothed in a warm coat of gray down with a halo of white down on its head, standing unsteadily on its disproportionately large yellow talons with heavy wings flopping droopily to each side, my first baby eagle looked sweet and vulnerable, not vicious and aggressive as I'd been led to expect. It reminded me of a little white-bonneted grandmother with piercing eyes.

"Little Jesus? Oh, you wouldn't!"

"I think it's very appropriate. Little Jesus came from the highest tree right out of the sky."

And despite my protests the name stuck.

A corner of the cabin was reserved for the new member of our household. It had to be a very wide corner as eagles can fire their droppings over six feet: though we spread many newspapers on the floor Little Jesus was very accurate in keeping his nest box clean and unfouled, to the detriment of our part of the cabin. Daily I had to weigh, measure, and chart his feather growth, to be used as a comparison in aging the banded nestlings.

At hatching, the eaglet's white down feathers offer little protection from the customary wet and cold weather in the nest and the parent bird must constantly brood her helpless young. After ten days, a warm gray down has pushed out all but a halo of white down on the head. At first the feet and the head of the eaglet seem grotesquely large. This is nature's plan:

an eagle must early in life have strong weapons with which to grasp and tear apart its food. At two months the one or two (seldom three) young begin bouncing up and down on the nest platform exercising their wings, and sometimes a young bird may make a premature flight out of the nest. Unless it can return to the nest it will starve: the adult birds will not feed it anywhere but in the nest. By the eleventh or twelfth week most eaglets make their first flight, usually with much exaggerated wing movement and an unstable landing, like a pilot bringing to earth his first plane.

I was responsible for feeding Little Jesus. The problem was not to get the food into him. No sooner was the meat held in front of his beak than it was swallowed in an instant. The problem was to catch enough for him.

Neither David nor Lionel were ardent fishermen. Rather than spend valuable time in jigging cod each day David decided to utilize an old Indian fish trap that was set up on the beach. The trap was a simple corral built of piled boulders. At high tide the sea perch floated in over the rocks. When the tide receded the fish were trapped behind the barrier of stones.

All I had to do was to visit the trap at low tide and pick up the perch for our eaglet. It sounds simple but not even Little Jesus screaming his shrill hunger pangs could make me pick up one of those slimy, goggle-eyed perch. David was not at all sympathetic. "Listen, Lyn, I love you, but you'll just have to learn to conquer your aversion to fish. Everything I study eats fish so you'll be exposed to fish for the rest of your life. Do you think a course with a psychiatrist would help?"

I had no intention of spending good money lying on a psychiatrist's couch, so I tried getting perch out of the trap by impaling each one on a stick, then waving both the fish and the spear in front of the eagle.

Neither of us like hunting but I was very glad when David soon shot a black bear a few yards from our cabin. The bear was put to good use. It provided Little Jesus daily with fresh meat. I concocted a dozen different recipes for our own dinners after our usual nineteen-hour day. We had bear roasts, bear stew, bear chops, bear steaks, bear liver, even chili con bear and bear hamburger. The skin was washed, salted, and packed away on the *Lady Rose* to be tanned.

David wanted to keep the skull for his scientific collection so he gave me the head of the bear in a box with instructions to put it in the fish trap for the crabs to clean out during the summer. Unfortunately I didn't weight it down, so the precious skull was lost when the box floated out to sea. David was a little mortified at my ignorance.

I had more success with my oyster garden. At low tide I could slosh my way to several of the nearby islets, collect oysters and mussels from more abundant sources, and plant them in front of the cabin for easier picking at mealtimes. At Barkley Sound our meals were more varied and better cooked than at any other time in our married life: as well as our staple diet of bear, we had oyster stew, clam bisque, buttered mussels, and seal liver. If I couldn't impress my husband with my practical talents I found the way to his heart was through his sweet tooth. David would start at sunrise with a chocolate pudding varied at times by rice or bread-and-butter custard. I cooked up to six desserts a day on the old wood stove.

At any one time my stove might have a custard on one burner, a boiled crow or another of Lionel's university specimens on the second, and a pair of underpants on the third. A considerable part of each day was spent collecting and chopping firewood to feed the insatiable hunger of the stove.

I had no intention of spending all my time in the kitchen so washing and cleaning were kept to a minimum. Washing became a matter of spreading the dirt evenly. The all-important priority was drying out one's wet clothes in the incessantly miserable weather.

David mounted a telescope in front of the cabin door trained on one of five nearby eagle nests and I spent as much time as possible making notes on the eagles that occupied these nests. David wanted to know what an eagle did all day long, where he roamed, when, and how far. As an eagle spent several hours just sitting on his perch staring into the water it became tedious keeping my eyes glued to the telescope waiting for some movement. And, needless to say, the very moment I took my eyes from the perched eagle, he would choose that moment to fly somewhere else and I'd lose him. In my spare time I prepared a course of high-school French which I was to teach later in the summer.

Once a day we'd check our eagle traps. David wanted to capture as many adult eagles as possible for banding and color marking. It was easy enough to band an eaglet in the nest, much more difficult to capture an eagle on the wing. The boys would spear a perch from our productive fish trap or jig for cod over the side of the boat (at times like these I'd find something else to do!). David then tied the fish to a log on the beach and arranged a few nylon nooses around it.

As an eagle is more of a scavenger than a hunter we hoped it wouldn't be too long before one of them got himself entangled in a noose. When an eagle was trapped the bird would be measured, weighed, and sexed, an aluminum band would be clamped to one leg and our own color markers applied. Some eagles we released had bright-colored streamers flying from one foot. Others had red or yellow or orange feathers imped in to replace a few of their own.

"Imping in" colored feathers for identification is a technique that has been used by falconers for thousands of years. The idea is to implant a previously prepainted feather as a substitute for one of the original feathers. A steel shaft with glue is inserted into the hollow shaft of the painted feather so that the steel shaft enters one end and protrudes from the other. The original feather is cut off to the same length as the painted one, then glue is put on the steel shaft protruding from the colored portion and inserted into the hollow shaft which is still attached to the bird. After a few seconds for drying the glue, you have a bird now plumage perfect except for one colored feather.

A feather, of course, is a dead structure, so it doesn't hurt the bird to cut it off or to have a steel shaft inserted into the feather. Depending on what wing or what side of the tail you insert the color in, and depending on how many you use, a unique series of color combinations for recognizing individual birds is created. Aluminum bands may be more permanent but the only way to identify a banded bird is to recapture the bird or be close enough to read the number on the band with a telescope—hence the advantage of imping the birds to identify them.

The shoreline of the islands in the sound was more interesting than the impenetrable forest area. Mink and deer swam between islands, families of mergansers scuttled across the surface of the water as fast as their motorboat legs could carry them; raccoons nosed among the tide pools for tasty sculpins or grubbed energetically in the mud flats at the extruded siphons of the buried clams; black bears turned over the rocks at the water's edge for crabs; and harbor seals with their pups poked their heads curiously from the kelp beds.

On one memorable occasion while David was out in the

boat with Lionel and Salal Joe, he spent many exciting hours observing the killing of a minke whale by seven killer whales. A pod of three adult males and two adult females each with a calf had chased a minke whale through a narrow channel into a shallow bay. Zero tide made the three possible exits from the bay unnavigable. The males attacked, while the females and calves remained about seventy-five yards away until the kill was complete. For the next three hours the men in the boat could only guess the details of the grisly underwater banquet. Every few seconds one or more killer whales surfaced. Twice only did the minke whale surface—vertically, with its head rising eight feet out of the water. Ten minutes later the unfortunate animal was seen alive for the last time.

Soon the entire bay was covered with a film of oil. Small pieces of flesh periodically rose to the surface. After one hour of eating, a six-foot white piece appeared, to be taken down by a male. Half an hour later a second large white object appeared, which was taken down by a female, only to reappear in the same place a few seconds later.

"Boy, I wish I had some of that meat," exclaimed David. Offhandedly Salal Joe remarked, "My scow is down the beach. I'll go and pick some up for you."

In his homemade fourteen-foot scow and armed only with a .22 rifle Joe headed across the bay. When he reached the area of surfacing whales, his speed and nonchalance decreased visibly. Without stopping the motor he reached over the side of the scow with his gaff and hooked a piece of meat sufficiently large that if he had not quickly let go of the gaff he would have been pulled overboard. Continuing across to the other side of the bay to gather up his courage, he approached the dinner table again, this time more slowly and cautiously. Grabbing the gaff with the meat still attached, he carefully towed it back across the bay. He had acquired a six-foot strip of the dead whale's back with the entire dorsal fin still intact. David then decided to take his chance among the feasting pod. He returned with ten pieces of the whale's tongue—the largest single section weighed ten pounds.

Next day Lionel found the remains of the victim about four and a half miles from the kill area and towed it to shore within walking distance at low tide from our cabin. Before it was necessary to peg our noses we studied and photographed the whale in detail. It was certainly the victim of the killer-whale attack. It lacked a dorsal fin and the tongue and flesh of the lower jaw were missing. From the torn anal opening the bloated intestine was protruding. But what amazed David the most was that the entire outer skin had been neatly ripped off, to expose the blubber layer and give the whale the appearance of a peeled orange. David guessed that the whale died from drowning. There had been very little blood in the area of the kill and no frenzied activity by the predators, just a systematic stripping of the victim's complete skin.

As the summer proceeded the carcass provided great scavenging for the eagles and great stimulus for our noses.

To learn about the food of bald eagles we hid near places where they fed. Five pairs of eagles with active nests lived near the dead whale. The first eagle to settle on the carcass was the female from the nest immediately above. Hardly had she started to eat than three more adults and a three-year-old eagle appeared. Within half an hour we could see twenty-seven eagles perched nearby. Only one or two eagles fed at one time. If another eagle came near, the one that was already eating would threaten it with low guttural sounds, a hunched neck, and raised wings which obviously meant "Get back."

The owners of the nest made no attempt to drive the other

eagles away. In fact, three other eagles sat in the nest tree, though well below the nest itself. Similarly, whenever we put seal carcasses near other nests, the eagles only defended a cone-shaped air space immediately above the nest. Though an eagle flying over a nest might be considered a threat, eagles flying below the nest apparently did not constitute a threat.

Sharing a large carcass seems to help the eagles to survive. A single eagle family couldn't use a whole carcass before it rotted or drifted to sea with the next tide, and it seemed nature's way to have the eagles share such temporary food supplies without wasting effort fighting about it.

The eagles were fortunate in Barkley Sound that summer. Many seal carcasses lined the beaches, owing to great on-slaughts on the harbor-seal population by hunters who received high prices for their pelts.

By field observations, and laboratory analysis of thousands of food remains collected in and below eagle nests, David learned that the eagle is an active predator during the breeding season, when it catches tiny fish, easily transported for the nestlings. But in the nonbreeding season, when the adult is freer to move about and free of family ties, it prefers to scavenge a meal rather than hunt.

As we all existed on less than five hours' sleep each night, we were glad to find that a popular feeding spot for the eagles was the eel-grass beds and tide pools only a few yards away from our bed. With the first light we were up with the birds for observation. As the tide marched down so did the crows, poking and prying under each newly exposed blade of eel grass and into the crevices of each pool. When the crows spotted something small enough to eat they gulped it down. Often they surprised fish, perhaps three to six inches long, too big for them to swallow in one gulp, then down would swoop one of the eagles from the surrounding trees to steal the fish amid caws of protest from the crows. This continued for the three or four hours of low tide, or until all the eaglets and adults were satisfied.

During much of the rest of the day the eagles watched the water from their favorite perches for balls of herring or other surfacing fish, which they swooped down to clutch in their talons. A small and easily swallowed fish is consumed in flight. This is easier said than done. With a small fish pinioned in the talons, in order to get the fish to his bill the eagle must swing his feet forward and simultaneously stretch his bill down, making almost a circle with his body, a very complicated maneuver in normal flight. In preparation he must slow down, almost stalling, and at the last minute tilt his body upward to gain elevation, swing his feet forward and reach down with his beak to take the fish. At this point he has lost nearly all his flying speed: he stalls and starts to drop. The only saving factor in making this clumsy transfer (particularly if he has meantime dropped the fish) is that his head is now pointing down to start him on a downhill glide to regain the forward speed he had lost.

During the time David has been studying eagles he has heard some fantastic tales about eagles that can carry off dogs, lambs, and children. But a male eagle with a wingspread of five to six and a half feet weighs only five to seven pounds, so most male eagles can lift no more than one and a half pounds at most. The female eagle, which is larger, with a wingspan of six to seven and a half feet, weighing eight to twelve pounds, can carry a maximum weight of four pounds. Though an eagle can lift a larger fish than it can carry, the weight can drag it into the water. It will then use its wings as oars, and row to shore dragging the fish along in its talons.

One day we noticed a large adult eagle swimming, and David fumbled for the movie camera.

"Lyn, grab his wing and flip him into the boat. But be sure to grab him before he grabs you with those talons."

I probably looked as horrified as I felt. But David was calm. "Go on. It's easy. You catch the eagle, I'll steer the boat and try to get it on film. Then we'll band it and release it."

I nervously prepared to follow instructions.

The eagle's method of swimming was fascinating. It used an overarm (or should that be overwing?) stroke. When the wings simultaneously hit the water in front it rapidly pulled its body forward, arching its body almost right out of the water. The movement was graceful, swift, and experienced. With these butterfly strokes, the envy of any Olympic swimmer, it twice outmaneuvered David's attempts to get me into a good position for catching it.

Once more David got bird, boat, and me into position. I reached out for its wing. Instantly it started to roll over and withdraw the wing, ready to strike out with its powerful talons.

"Look out, Lyn!" David yelled, at the same time gunning the boat so that I grasped at air, then fell flat on my face in the bottom of the boat.

Again we edged closer, while the eagle expertly swam a zigzag course to evade us.

"If you miss catching its wing as it rolls to strike then get back," David cautioned. "One more eagle isn't worth a gashed hand or arm—or worse."

With its massive yellow scimitar of a beak waving up and down I knew it was protesting our presence. Again my arm shot out and my hand gripped the long flight-feather shafts.

In one continuous motion, two powerful wings, two slashing talons, and a snapping beak burst from the water and landed on the boxes in the boat.

"Quickly!" David yelled. "Push its back down, pin the feet and talons underneath. Careful of the wing butts. Then carefully get hold of the legs so the talons don't get hold of you."

As my hands were totally involved I had only my head to distract the wet and angry eagle. By pulling its wing I kept it off balance so that the dreaded talons were kept busy clutching the boxes for a footing. Simultaneously I slid my hand up its tail and pressed down until I had the tail feathers and both legs firmly gripped in my fingers. Then slowly the eagle teetered over helplessly onto its belly, the dangerous talons safely secured, the powerful wing butts tucked in at the sides, and the massive beak pointing forward out of harm's reach. Click, click, click—the film ran out. David gave a sigh of relief and so did I. I think the eagle did too.

Swimming is a relatively common activity for bald eagles. But catching a swimming eagle was an entirely new event to me. With the eagle now in hand we measured it—a full seven-and-a-half-foot wingspan and a weight of eleven and a half pounds, a larger-than-usual adult female.

The next day I achieved another first when I climbed into an eagle's nest. It had been a wet, drizzly morning with the clouds hanging on the treetops. Leaving our cameras behind in the cabin we headed by boat for nest 137. When we reached the island that contained that particular nest, David dropped Lionel and me on the beach while he left to continue his boat survey of the eagles' habitat.

In record time Lionel reached the top of the tree, and from

somewhere in the tangle of branches above, his muffled voice called down to me: "Hey, Lyn, there are two eaglets in this nest. Come on up and have a look."

"You're crazy! You know I hate heights. Is this another tree easy enough for your grandmother to climb?"

"Seriously, Lyn, I think you can make this one. With a bowline around your waist you can't fall. Let's surprise David when he comes back and finds you sitting in the nest with two eagles."

"How big are the eagles?" I asked dubiously.

"They'll be ready to leave the nest any day. But they're surprisingly tame," he quickly added.

"Well . . . I don't imagine I'll get far."

Lionel's rope dangling down from the lowest branch started to twitch, a foot, then two legs, then a backside appeared through the snarl of branches. Then the rest of Lionel appeared. He tied me to him and we began the ascent.

Cautiously I worked my way up, over, under and through the slippery branches. With the bowline around my waist, I finally broke through the canopy of leaves to a four-foot-wide nest wherein sat Lionel and two fully grown eaglets.

Some eagle nests are large enough to hold a dozen people. This one was the smallest I had ever seen, cradled by three branches but with barely enough room for two eagles, let alone two people. Getting into the nest made me wish I hadn't started up the tree: I had to lean backward and climb over the protruding rim of the nest. For someone who fears heights, this wasn't easy. Fortunately I had the help of Lionel's reassuring rope, and happily the two eaglets also cooperated delightfully, and acted as if they received human visitors every day.

In a moment the wind blew some of the clouds away and

the view was superb. I hoped David would hurry back: I could hardly wait to see his expression when he saw his wife 150 feet up a Douglas fir, with an eaglet perched contentedly on either side of her.

The sun began to shine through the clouds and I persuaded Lionel to leave the measuring of the eagles until David came, so that we would record it on film. Two hours later we heard the reverberation of the outboard motor as the little dinghy slapped over the waves toward the beach. Gleefully, I waved my jacket. Our tree was slightly taller than those around it, so a girl furiously waving a red ski jacket from the nest platform was easily visible: David was so startled he drove the boat right up on the beach.

Two people and two eagles in a nest as small as this one made for cramped quarters. Three people and two eagles would be impossible, so Lionel climbed down to meet David. But, since it had been overcast when we set out and we hadn't brought the camera equipment that day, Lionel took the boat to go back to the cabin for it, while David climbed up to join me in the nest.

The wind intensified as the sunshine increased. Carefully David and I rearranged our positions. His legs hung over the edge of the nest, dangling high above the sea. More cautiously, I sat in the center of the nest, with one eaglet on my lap; the other nestled between us.

Two, and then three hours passed. I prayed for the sun to stay with us until the cameras arrived, while David lamented the fact that we hadn't brought the camera gear in the first place.

Another two hours dragged by. What was Lionel doing? We thought of all the things that might have prevented a quick return. Sitting so long in a cramped position was getting uncomfortable. We nibbled at our only two chocolate bars, and yet another hour went by.

Our impatience changed to worry: had Lionel had motor trouble? Was he stranded on some island? And we on another? Had he met a worse fate, on a now white-capped sea? We remembered that Lionel couldn't swim.

We decided it was wiser for us to stay in the vantage point of the nest. Besides, I didn't want to have to climb this tree a second time for the film sequence. The eaglets were now so used to our company that they went to sleep, one with its huge beak and head across my leg. The sun was dropping to the horizon, and the wind was lessening.

David and I were not the only ones to be anxious and hungry at the end of the ninth hour. A loud chittering close by caused both eaglets to look up. Overhead, one of the parent birds was bringing the evening meal. Both eaglets stood up, stretched their wings, and nearly fell out of the crowded nest. We knew they had to be fed, and even if Lionel did come, the light was by now too dim for photography. So David banded the eaglets and we climbed down from the tree to prepare a fire and leanto shelter for the evening.

Before David had the fire started, the adult eagle had taken food to the nest for her young, and just as our makeshift camp was ready, we heard a motor. Lionel had arrived with the cameras. As we feared, he had had trouble with the motor. To get it repaired, he had walked eight miles at low tide around the shores of Vancouver Island to a logging camp. We were relieved that he was safe, but it was too dark to use the cameras.

Passing by the nest three days later, we saw both parent eagles circling the nest. Then, as we watched, the young male launched himself into the air on his first flight. Out toward the boat he came. Immediately above us he turned left, then headed for shore, a hundred yards away. Awkwardly he approached the trees, his wings back-paddling furiously to help slow his speed. Thump, shuffle, shuffle: he landed, and folded his six-and-a-half-foot wing span. He was bigger than his father, but still wearing the brown feathers of a juvenile. His sister stayed in the nest.

She might not fly for another day or two. The female's growth rate is usually slower than that of the male. We kept watching them. Where would they be going? Would they find enough food? Would they be back here again in five years wearing the white head and tail feathers of adult bald eagles ready to rear their own young? We wondered and wished them well.

Before David's study the bald eagle was generally considered a year-round coastal resident. Over the two hundred square miles of southwestern British Columbia regularly covered on his census flights, David had documented definite seasonal changes in numbers.

Egg laying begins in the first week of March while the nearby hills are still shrouded in snow. After approximately thirty-five days of incubation, probably mostly by the female, the eggs hatch. Peak hatching occurs during the last week of April. By the middle of July, over fifty percent of the young have fledged. For the first few days after the fledging the young may return to the nest for food. But very quickly their center of operation shifts. The family unit stays together but changes its area of operation daily. Thus the migration starts, slowly but surely. Once started, the exodus is rapid. By the middle of August less than fifteen percent of the breeding population remains. Emigration is completed by the first week of September. Over the fifteen hundred square miles of the Gulf Islands there is a virtual absence of eagles between early

September and the end of October. The nesting on the outer west coast of Vancouver Island and Barkley Sound is about two weeks later than that in the more protected Gulf Islands between Vancouver Island and the mainland.

The return of the eagles occurs in two waves—the first, composed mostly of adults, arrives about the third week of October. The second wave, about twice the magnitude of the first, and composed equally of adults and immatures, arrives about the first two weeks of January. Many of these birds are undoubtedly breeding adults returning to their home range to winter after a very brief post-nesting migration, but the other adults and immatures present seem to be only temporarily using this area. Just after the peak population is reached in mid-January and just before or perhaps connected with the breeding season, the adult segment of the population begins to decline and redistribute. After territories have been established and nests repaired, mating is soon followed by egg laying to complete the cycle.

One of David's problems was to find out where exactly the eagles migrated and what purpose was served by this short period of migration. He suspected that they traveled northward to follow the annual run up the rivers of the spawning salmon. In a few hundred yards of river, David had once counted more than a hundred bald eagles feeding, being heckled by the crows and gulls, resting with distended crops and waiting to fill up again.

Leaving David and Lionel to delve into these problems and to classify all the plastic bags of rotting food remains I had collected under the nest trees, I returned to Vancouver to teach summer-school French and to prepare for the fall term.

4

Sam Goes to Jail

FLOPPING OUT THE WINDOW OF OUR ONE-TON CHEVROLET
camper, Sam thoroughly enjoyed the drive south on Highway
99. At the border, the U.S. customs officer apparently didn't
even notice a flipper waving idly in the air. We had gone to
a lot of trouble to get the necessary permits to take a seal on
a holiday, so we were a little perturbed when the man took a
quick look at Sam, sitting on the seat between us, and then
seemed to pretend he hadn't seen him. We were a few yards
into the United States when David abruptly turned the car
around and returned to Canada.

"If we don't register Sam going out of the country, we
might run into trouble getting him back into Canada again.
They might think we bought him in the States."

Sam the seal was duly registered as a Canadian citizen.

David had always been interested in educating the public
to the values of wildlife. In the month between his eagle
studies and his return to the university for the fall session, he
wanted to visit zoos and aquariums down the west coast as
far south as the Mexican border. It was the first time I had ever

accompanied a sea lion on a camping trip—and this was to be Sam's greatest adventure.

There was plenty of room for the three of us in the front seat. When Sam wasn't ripping up the route map or leaning out the window he lay on his back nibbling itches on his tummy. We bought gas only at those stations that didn't mind giving Sam a wash with their water hose.

At night Sam slept in a steel cage attached to the camper's side. The first night was noisy because Sam's movements shook the camper. The rattling sides of the cage just below our ear kept us in intimate contact with his activities. The second night in Oregon was peaceful as we unbolted the cage from the truck and set it on the ground.

At seven in the morning I was wakened by sun streaming through the curtains. I picked up a breakfast herring for Sam and slipped outside. The cage was empty.

"David," I shouted, "Sam's gone!"

"You're kidding," he answered, but he rushed outside, took one look at the deserted cage and shot down the highway in his underwear. Meanwhile I raced down to the Santiam River, wondering if Sam had felt the lure of the water. I ran along the bank calling out to every camper I saw, "I've lost a seal. Have you seen him?"

Can you imagine a woman in nightgown and curlers hunting a seal down by a river at seven in the morning?

We had been camped right beside the highway. If Sam had decided to try to cross it, his chances of survival were slim. On the other hand, if he had reached the river would he swim upstream or down? I tried to put myself in the place of a seal. Would he drift downstream easily with the current? Or would he prefer the challenge of fighting the rapids and swim upstream? I stood on the bank and stared for several minutes at a flat blob that looked like a head. The blob moved slowly upstream. I called, and it moved a little closer. It must be Sam. I ran back toward the camper, bumping into David, who was still half naked.

"David," I yelled. "I've seen Sam. He's in the river and swimming upstream."

"Good. Go back to the camper and bring my snorkel and half a dozen..."

He was already dashing off to the river and I could only guess at the rest of his sentence. Half dozen what? I grabbed a collar, leash, string, and half a dozen fish, then raced back to the river to meet David disconsolately returning up the path, minus Sam.

"That wasn't Sam. It was just a land otter. They both have similar heads."

We returned to the camper and dressed hurriedly. Now to convince people we really had lost our seal.

"We'd better get on the phone to the newspaper and radio and television stations so as many people can be alerted as possible," David said. "Of course he could be lying low in any of the bushes around these farms. Let's call at the first farmhouse and ask them to spread the word around."

After some consternation expressed at the door, the first farmer's wife took a chance and let us in. We phoned the Portland Zoo, Washington State Police and Louis's Boats and Motors in Albany, Oregon, as well as the local news media. We stopped cars on the highway and asked them if they had seen a seal. We phoned through to alert the dam on the Columbia River in case Sam was heading to the ocean.

We drove to Albany to borrow a boat and motor. With a ten-horsepower Mercury on a flat-bottomed boat we set off downstream on the Santiam River on its way to join the Willamette River. Our eyes strained at the banks and we searched every patch of water, every tiny island. We used motor and oars alternately as the river swirled and eddied.

A ferry boatman crossing our path hadn't seen Sam and his expression indicated that he thought he never would. Families with summer cottages along the river promised they'd watch for a passing seal. And so it went for twenty miles, until the river became so wide it was impossible to scan the banks. We hauled out our boat on a sandbar near the little town of Independence, where a truck was unloading gravel. The police patrol car drove us back to our campsite near Santiam.

We checked the area again but found no sight of pointed nose or tiny ears or saucy whiskers breaking the sun-streaked surface of the river. We looked behind every bush, parted every clump of grass, calling "Sammy! Good Sam! Nice herring, Sam!" It seemed silly enough to lose a seal but now suddenly it began to seem even sillier to expect to get him back.

That night we left the door open with a couple of herring on the mat as a welcome-home present. They were still there in the morning.

Our next plan was to hike along the bank, via the blackberry bushes, for a few miles upstream then a few miles downstream. Although we were scratched and gouged by stinging nettles and blackberry bushes we continued to call "Sammy!" to every piece of wood floating in the water and every movement in the bushes. Exhausted, we returned through fields of carrots, potatoes and corn to the camper. Some newspaper reporters were waiting to interview us. They had no encouragement to give us despite the fact that Sam's disappearance had made coast-to-coast news.

We had done all we could. Desolate, and resigning ourselves to a life without Sam, we continued south. During the next three weeks I tried not to think of seals or sea lions as we walked around every zoo, aquarium and marine center in Oregon and California through to Mexico. Every few days David would call back to the Portland Zoo for news from the zoo veterinarian, Dr. Maberry. David was still hopeful. I felt sorry for him: I never expected to see or hear of Sam again.

Then one morning in Los Angeles, while we were staying with one of the whale trainers from Marineland of the Pacific, there came a call for David from Portland Zoo. It was Dr. Maberry and he had found Sam. Impossible but true.

After more than a week Huck Finning it in the river, Sam had crawled out of the water on that same stretch of sand near Independence. The same truck driver was there and he called the same policeman who had given us a lift. The policeman enlisted a posse of two schoolboys and they set off to capture a seal sunning himself on the beach.

When they arrived at the beach Sam was back in the water. Armed with salmon-landing nets they followed him in a boat. For three hours they whistled, shouted, and tried to coax Sam into a net. But playfully Sam dodged around and under the boat, obviously delighted by the game.

Finally a boat arrived on the scene with a couple of newspaper photographers. They tried to attract Sam's attention long enough to get pictures while the boys glided up from the rear and landed him in the net. I am sure if Sam had been a wild seal they never would have caught him.

The *Oregon Journal* finished the story: ". . . What happened to Sam was downright humiliating. First they fished him out of the Willamette River and threw him into jail. Then they took him to the zoo and put him in the monkey house. And Sam isn't even a monkey. He's a seal."

A week later we arrived at the Portland Zoo to collect our

runaway—now nicknamed in all the papers as Slippery Sam.

As soon as we stepped in the Primate House he recognized us. Although noticeably thinner than when we'd last seen him, he'd obviously been catching food in the wild—though not enough. We were delighted that he was pleased to see us. We tied a leash to his choke collar and led him through the rest of the zoo.

The effect on the other caged animals was fascinating. If a human being passed their cages they were unconcerned. But to see a seal flippering along was too much for one big male lion, who left his lioness in the very middle of mating to stare with interest at Sam as he passed. I will never forget the expression on the face of that lion—or of the lioness as her suitor abandoned her.

There were other occasions on our return home when we thought we'd lost Sam again. Once we were visiting our friends the Slade family, to discuss buying some of their furniture that we hoped might be sealproof, when Ray Slade casually remarked, "By the way how is Sam? Fully recovered from his adventure?"

Immediately David and I looked at each other in horror. At the same instant we realized we had not closed the back gate to our yard. Without stopping for explanations we dropped everything and bolted like lightning to the car. We roared through stoplights, broke all speed limits, and skidded breathlessly into our house—to find Sam lying on the grass beside the open gate just as we had left him. We lolled on the grass beside him until our nerves had recovered enough to enter the house and phone back explanations to the Slades.

One dark, squally night later in the fall Sam managed to climb over the fence and flipper his way down the street. It was almost midnight when we returned home to find him gone.

There was no sleep for us that night. We drove up one street and down another all over the neighborhood looking for him, desperate at the thought that he could so easily have been run over in the rain and fog.

Unfortunately, Sam had picked the worst night of the year to go for a stroll. It was Grey Cup Night, Canada's annual night of football insanity, and Vancouver was celebrating the event to the full. David phoned all the radio stations but they all thought he was stringing them a line as a Grey Cup joke. The typical reaction was "So you have lost your pet seal, have you? He just got out of bed and walked away, did he?" followed by raucous laughter.

One or two of the radio stations allowed David to tell his story over the air himself: that way, if anybody sounded idiotic it wouldn't be the station's fault.

As soon as it was light David drove around the streets again. We phoned all the neighbors, although we fully realized no one liked being awakened so early on a Sunday morning. In the constant rain, nobody wanted to form a posse to locate an elusive seal. Finally Dorothy Stewart, the kind of neighbor who, with five children to worry about, could always find time to help somebody else, braved the elements to hike through the nearby Indian reserve in search of Sam. It was Dorothy who found him, bounding along the side of the road no more than a block from home.

On my first Thanksgiving in British Columbia we nearly lost Sam for another reason. Sam had started early to have all the "childhood diseases." He seemed particularly inclined to watery eyes, coughs, colds, and pneumonia. Wild fur seals, born on the snow-bleached purified rocks of the Bering Sea and spending most of their life in the sea, are a unique species that have developed no internal resistance to diseases of civilization. To build up his resistance and good health we experimented with all kinds of vitamin additives and antibiotics.

After an enormous dinner of turkey and all the trimmings, David and I were relaxing in the living room when Sam was suddenly taken ill. He regurgitated all his own Thanksgiving dinner of smelt, then quickly became delirious, twisting into most peculiar positions. We moved him to the bathroom floor, where he writhed in alternate circles and lunges. He lay inert —paralyzed and pitiful. Trying every injection and remedy he could think of, David alternated between the bathroom and the drugstore while I, feeling simply helpless, whispered into his ear, "You can't die, Sammy," trying to will him to live. After eight hours of futile attempts to revive him, David forced me away from my post. "There's nothing else we can do. Leave him."

It was only then that I realized how much Sam meant to our family. We both left him for dead.

Next morning David opened the bathroom door to see Sam alive and with the power of movement regained. It seemed miraculous. Since then Sam has never been seriously ill for as much as a day.

I tried to make it a habit to jot down notes on Sam's daily doings. Sometimes this got me into hot water—literally. One evening while I was filling the tub with hot water for a bath I had the sudden urge to rush to the kitchen to grab a pencil and describe some of Sam's amusing poses as he moved around the house. At that particular moment he was stretching out on the toilet seat with his head and neck stretching vertically up the wall, and rolling his golf-ball eyes. No sooner had I written one word than I heard Splash! This could only mean one thing. Sam had splashed from the toilet seat into the hot bath water. Thinking he would be scalded despite his insulating fur, I dropped my pencil and raced back into the bathroom. Surprisingly, Sam wasn't even trying to get out. I thought David would be furious because I had allowed Sam near a dangerous situation. But he merely asked, "Did Sam look worried?" In relief I answered "No" and Sam casually flipped out, splashing water all through the house.

In play, Sam was a hilarious attention seeker. With mouth wide open, a smile on his face, and making little gasping sounds, he dashed from his dog-kennel shelter to lunge at me whenever I passed. He loved being chased almost as much as chasing others. Of course such play among wild seals is really a rehearsal for the more serious challenges with older bulls later in life.

In the house he would fight with all flippers to climb up on my knee and nudge his nose against my cheek and encircle my face with his tough, tickling whiskers. He liked nothing better than to roll and romp on our bed. Incidentally, it is amazing that animals accustomed to a rigorous existence in the wild always choose the warmest, most comfortable bed inside the house for their territory.

When we had guests Sam was always the life of the party. He always studied each guest, picked out the one who was studiously looking the other way in the hope that Sam would do the same and flopped over to him with his characteristic quick lunge. Our Aunt Helen is usually the poor victim; in fact she has spent many evenings behind the closed door of our bathroom. Her visits have lessened considerably of late, come to think of it.

When I recalled my own school years I realized it wasn't the facts, skills, or the blanks I filled in on test papers that I remembered. No, it was the associations I had with my classmates and teachers in group activities, the interests that were stimulated or developed; in short, the involvement I felt in the school was the important factor in learning. As a teacher I have always tried to involve the students in wanting to learn because what they were doing was interesting. Sam provided instant motivation.

Sam came to school for the first time to demonstrate migration patterns. On Sam the Seal Day excitement mounted during the hours before his visit. Nobody could concentrate on anything but Sam's expected arrival at 2:25 P.M.

On the dot of 2:25 the door opened. As David came in he was totally ignored—all eyes were focused at floor level as Sam slip-slopped up to the teacher's table, and soon he was surrounded by eager children.

In response to a fishy bribe, Sam flopped around giving the girls rather damp kisses. He obviously loved the attention he was getting, and the children loved his sense of fun.

We soon learned one important thing about using animals in the classroom to stimulate learning. Never try to talk about the animals while the animal is absorbing the children's attention. A living creature in the classroom is too much competition for any teacher to fight. So, while David talked to the class about migration patterns and in particular about how fur seals migrate annually from the Pribilof Islands in the Bering Sea to the waters off California, I enticed Sam away, down the corridor, to visit a few more eagerly waiting classes.

Just before the bell I tried to sneak him off in the car. But rumor travels fast, and one minute after the bell it seemed that the whole school was crowding around the car windows for a last look at the hero of the day. After that Sam became a regular visitor to the school. He illustrated our science and our social-studies lessons and provided a great stimulus for creative writing.

Mary Ellen described Sam's first visit this way:

Mrs. Hancock, our teacher, has a pet fur seal named Sam. He is a lovable animal with dark shiny fur and big brown eyes like golf balls. Boy! Does he get into mischief! He does all kinds of tricks, such as shaking flippers, rolling over and kissing—all for a piece of herring. Our vice-principal took pictures of him in various poses and we added humorous captions to them for Open House. When Sammy was ready to leave, Kim the class representative shook hands with him and said goodbye.

When Sam wasn't visiting schools, schools were visiting him. After four moves in four months we had settled in a small four-room bungalow near the university—and sometimes too near the local elementary school. Before school, during recess, at lunch time, and after school as well as on weekends we ran a constant baby-sitting service for the neighborhood. Children flocked to our backyard. Parents brought their weekend visitors. David had to alter the gate to a swing type so that when everybody left, the gate would automatically close.

Our backyard was the perfect setting for Sam's playful antics—it was seventy feet long and enclosed by a picket fence, with a front gate that led to the street and a back gate leading to the alley. Daily, children entered through each gate.

The action started when Sam first tested each youngster to see which one would run. Those who knew him called his bluff and stood their ground, and they were duly ignored. But timid newcomers were relentlessly pursued as they fled in hysterical fright to the fences. There Sam had them trapped. The gates at either end of the yard opened inward. Anyone leaving by the gate had to take a couple of steps closer to Sam, in order to squirm around the gate and out of the yard. By the bristly twitch of Sam's whiskers and the gleam in his eye we knew that his moment of greatest triumph came when from the middle of the yard he was able to command at the same time two nervous youngsters, one at each end of the yard, too panicky to move. He never hurt any of his victims but he did a pretty good job of scaring them.

Soon Sam had to share his yard with other animals as well as with people. The sharing began with the arrival of two little falcons.

Almost every June, David has flown to the Queen Charlotte Islands to study the decline of the peregrine falcon population. The falcon, fastest bird in the world, clocked at 200 mph in a steep dive, was formerly found all over the world, but now the misty, beautiful Queen Charlotte Islands are one of the last strongholds for this endangered species. David originally had been a falconer. Now his main interest was to maintain pairs of falcons for breeding, with the eventual aim of releasing falcons bred in captivity into the wild, and to provide a source of falcons for falconers to eliminate harvesting of the fast disappearing wild ones.

Falcons are used to keeping still in their eyries on narrow ledges halfway up steep cliffs, so it was an easy matter to fix a nest for our two soft downies. I merely nailed a shelf to the wall of the garage. Gal and Charlotte were screamers: every time anyone approached they opened their beaks and let forth such squalls that I had to pop in some mashed-up chicken heads to keep them quiet. Even then they continued to yell, though admittedly the cries were muffled until the food reached their crops, after which their raucous cries resumed at full strength.

One morning I awoke at six to see Sam chasing Gal around the lawn. The falcon must have been trying out her growing adult feathers by doing her jumping exercises on the ledge and had landed off her perch and onto the lawn. Once again attired in nightie and curlers and armed with a salmon landing net I called Sam away and chased the falcon around the lawn until I retrieved her. Sam looked most annoyed.

Later on in the spring David added to my backyard menagerie two gray-downed month-old eaglets. Little Jesus, now almost a year old, resided at the folks' farm in Victoria. The little eagles as well as the falcons were ideal subjects for a school study of the feather and its development. I named them Tom and Jerry.

Though babies, they were huge, and we couldn't nail *their* nest to the garage wall. So David mounted a wide, grass-filled box on top of a chest and set it up in the middle of the yard beside Sam's swimming pool.

For a couple of weeks I took Tom and Jerry to school with me every day. Considerable space had to be left around them. There was no way to know in what direction the birds would be pointing when they had to excrete. They could thoroughly whitewash a hapless spectator at a distance of more than six feet.

One Sunday visit to a friend was rudely interrupted when a neighbor called to report that Sam had climbed into the eagles' nest and ousted the two eaglets. He was reportedly sitting on the tea chest as self-styled King of the Backyard. Tom and Jerry were waddling indignantly beneath him on the grass. I couldn't help laughing when I rushed home and saw the astonished expression on the faces of those eaglets.

After the many months it had taken to get used to cutting up herring, I was surprised that it took only a few days to accustom myself to chopping chicken. Mrs. Clemons, one of the supervisors from the school board, offered to help me with the chores. After describing the grisly preparation of the chicken heads I felt sure she'd change her mind. Despite my warnings that chicken heads had to be mashed with a hammer and scissored up in small pieces soft enough for baby crops, that she might be nauseated by the effects of mashed beaks and bashed eyeballs or the accusing look of dead hammered chicken faces, she didn't falter in her resolve. On the other hand, neither did she volunteer a second time.

5

Sea Parrots on Solander

It was a day of great excitement in the zoological world. Press cameras clicked, curators and keepers strutted proudly. The focus of all this attention was a black, bug-eyed puffin chick, which, after fifty days of living in the darkness of his fiberglass burrow, had emerged into the bright, busy world of the New York Zoological Society's Aquatic Bird House. He was the first alcid—or true diving sea bird—to be bred in captivity, anywhere in the world. This was a proud moment for William Conway, Director of the New York Bronx Zoo and instigator of our Solander expedition, our first animal-collecting experience. This celebrated chick now basking in the limelight was the offspring of our trip to collect puffins on Solander Island, a tiny point of jagged rock off the northwest coast of Vancouver Island.

David's studies and experiences, I'm happy to say, are all involved with live animals, unlike those of some of his university and museum friends who collect dead specimens. As soon as my teaching term and David's eagle studies were completed for the year, we prepared for our expedition to Solander Island.

Little is known about the puffins, or indeed about any of the alcids in British Columbia. David wanted to observe the local birds and compare their habits with what was already known about puffins in other parts of the world.

The flight from Vancouver in a commercially operated Goose float plane was spectacular. The green mosaic of the Gulf Islands gave way to the jagged snow-clad mountains splashed with emerald glacial lakes of central Vancouver Island. Below us, much of the land lay denuded—raped and scarred by logging operations. In an hour we had our first glimpse of the long fingers of the Pacific Ocean which pierced the mountains of Strathcona Park.

My ears began such a hammering that I realized we must be descending rapidly. I swallowed hard and wagged my jaw quickly sideways to quell the pain in my ears caused by the increasing air pressure as we docked at Zeballos, a mining and logging village at the head of Esperanza Inlet, on the west coast of Vancouver Island. One of David's flying acquaintances pulled us into place and helped us to transfer our baggage to the Cessna 180 for the ten-minute flip to Kyuquot, our jumping-off place for Solander Island.

Kyuquot was the prettiest fishing village I had yet seen on the west coast. It is a crescent-shaped island, closely hugging the head of a fjord. Two stores and a few wooden houses lined the inner part of the bay. Two of the houses had gardens with roses climbing over white-painted fences, a sign of civilization I hadn't expected to find in this part of the country.

On this late Saturday afternoon, boats of all kinds were coming and going in the tiny harbor. Fishermen were unloading the day's catch, carrying water, piling stores aboard, and fueling up. Off David strode to ask the Fisheries Officer if

there was a fish boat which could drop us off on Solander the
following day.

I sat on our baggage piled awkwardly on the float and
pondered: Where would we sleep tonight? I envisaged camping for the night on the float beside our six hundred pounds of
gear—wooden shipping crates for the birds, pens for holding
puffins, three weeks' supply of food, gas, water, stove, clothing,
sleeping gear, cameras, and tent.

Within five minutes David returned, his eyes alight and his
manner enthusiastic. "Doug Henderson, the Fisheries Inspector,
has arranged for a fisherman to take us to Solander on Monday
on his way north to Winter Harbor. We can sleep on the
floor of the store tonight and fill in tomorrow around here. So
let's get these boxes into that shed and then have a coffee with
the lady in the house over there—the one with the roses!"

Leaving the baggage in the open shed on the dock, we took
our sleeping bags and cameras to the storeroom above the
second shop at the end of "Main Street"—a narrow boardwalk raised above the brush and garbage that edged the beach.
We then called on Lucy Kayra.

Lucy and her husband Esco lived in an unusually modern
and fully equipped house overlooking the harbor. Esco was a
successful fisherman and Lucy an excellent baker of homemade bread. Coincidentally she also had a pet seal, a female
harbor seal she called Charlie, who roamed loose in the bay.
The whole of Kyuquot was Charlie's playground. She swam
among the fish boats, flopped out on the floats or the beaches,
and waddled to the Kayras' front gate when she was hungry.
Notices were prominently displayed around the cove to warn
visitors of her special place in the community. As an extra pre-
caution Esco painted her head white to set her apart as a pet
seal, and deter hunters who might fancy her pelt.

With a seal in common, we spent the next few hours chatting with the Kayras over a freshly baked loaf of home-made bread. I was keen to know how they had originally acquired this free-roaming seal. And how did Charlie get her rather misleading name?

"Well, we had our nephew staying with us at the time of Charlie's arrival and we called her after him," Lucy said. "A local hunter harvesting seals for their pelts performed a Caesarean operation on a pregnant female and brought Charlie into the world. We fed her a formula of milk and cod-liver oil. At first she was so small that she slept in the bread box and swam in the bathtub."

With the sea at her front gate it seemed that the orphan had the best of both worlds—round-the-clock care and atten-tion from Lucy in the house for the first two weeks and later an introduction to the sea when Esco built her a floating crate in the salt water beside the float.

As with all animals, weaning is the difficult period, but Lucy persevered. With patience and intuition she overcame the initial problems of diarrhea and the pup's refusal to take solid food. At last after a month Charlie was swallowing strips of salmon.

But the big problem was to teach her how to catch fish for herself. As with most wild animals taken in a domestic situation, this is something which doesn't come naturally. Charlie had been so used to recognizing food as orange sockeye salmon that it was almost impossible to teach her to eat any-thing else. Lucy told us how they handled the problem.

"To accustom Charlie to recognize silver movement in the water we put a herring on a string and got her to run after it. It wasn't until fall that she caught her first fish—a dog salmon. She brought it home and dropped it at my feet: she wanted me to cut it up. Soon after that she brought home an octopus. And on Christmas Day she brought us a present of a ling cod."

The first few times Charlie left the cove for any length of time, the Kayras were on tenterhooks awaiting her return. How was a tame seal going to cope with the hazards of the wild?

On one occasion an Indian fisherman from the nearby village on Aktis Island recognized Charlie near a pod of killer whales. He returned to Kyuquot with the seal swimming beside his boat and proudly announced that he had saved her from the killer whales. She had spent three days beside his boat and then followed him into the harbor.

Esco frequently took Charlie out on abalone diving trips in the speedboat. On one occasion when they were more then ten miles from Kyuquot Charlie jumped out of his lap and dived overboard. When the time came to leave Esco couldn't find her and returned sadly to Kyuquot.

"You'd think he'd lost his own kid," Lucy said. "Next morning at four o'clock he left in the *Courageous* to look for Charlie—she's more familiar with the sound of the big troller's motor than the speed boat's. Still no Charlie. At last three days later Charlie flopped through the gate after finding her way home—through ten miles of strange territory!"

Raising an orphan seal saves its life but it poses the problem of adapting it to the wild, especially mating it with its own kind. When Charlie was in heat Esco took her to a nearby seal breeding rock. Unfortunately she showed no interest in the males of the colony.

Sherry, their teenage daughter, had her own ideas on the subject. "Gee, I think Daddy thinks more of Charlie than me. He wouldn't take me and leave me with a bunch of males."

The first year the Kayras spent the winter in Victoria,

Charlie seemed to guess that the family was leaving for a long time. As Lucy and the children piled their suitcases on the dock and climbed into the float plane, Charlie followed them onto the pontoons of the plane. The following spring when they returned to Kyuquot there was Charlie on the dock to meet them as if once again she had known their plans.

Both Esco and Lucy were politely incredulous that we should be planning to spend the following three weeks on Solander Island. It seemed that although Solander was only forty miles away few if any of the local people ever visited it.

"Can't think why you'd want to go there. It's just a rock with a bunch of birds," they laughingly derided as they wished us goodnight.

I was glad to throw my sleeping bag down on the storeroom floor beside the haberdashery and go to sleep, wondering why David remained enthusiastic about Solander when the locals thought so little of it.

Next morning we hiked the trail around the narrow island. The sunny smiles and friendly greetings of everyone we met matched the glorious weather. Charlie flopped out on the dock and reached for salmon strips offered by the Kayra children. He swam with them, bit their feet, and pulled at their oars. What a perfect way to own a pet! Free and in the wild, yet tame enough to wriggle up the basement stairs when he wants a fish.

Suddenly all the children started to shout, "The barge is coming, the barge is coming!" Dogs, children, and adults emerged from houses that surrounded the bay. The barge bringing in the week's supplies was one of the highlights of the Kyuquot week.

"Let's go down and help carry up the boxes to the store, Lyn," suggested David. "They say that Wally Arnett is piloting it today. It'll give us a chance to meet him before he takes us to Solander."

We had seen Wally's boat the *Audrey H* tied up to the dock. Obviously no fisherman capable of catching five hundred to a thousand dollars' worth of fish per day would take time off to drive the barge or offer to take a couple of nutty bird-watchers out to a forbidding rock, so we approached the only available skipper with some apprehension. Would he be competent, friendly, and willing?

I liked Wally the minute I saw him. He had a kind face, blue sparkling eyes, and gray hair. He was middle-aged, slim, and athletic, with a wiry body hardened by life at sea.

Lucy and the frail-looking lady from the store came to help us tote the boxes from the barge to the warehouse. The heaviest boxes were loaded on a wheeled cart, then hauled up from the dock by cable.

"See you at seven A.M. tomorrow by the *Audrey H*," Wally called as the last boxes were unloaded.

Down at the Co-op, Esco was unloading his seven hundred pounds of Coho salmon from the decks of the *Courageous*, a troller as neat and fully equipped as his house. He paused as we came along.

"Real rough outside, Dave. Cape Cook is smoking. Doubt if you'll get onto Solander tomorrow in that lump. Who do you have to help you ashore?"

Dave's answer was prompt. "My wife, Lyn." And, as Esco looked questioningly, David quickly added, "You don't know these Australians. They're a tough breed."

Esco still looked as dubious as I felt. "I doubt you'll ever land that light skiff of Wally's in the six- or eight-foot swells I saw out there today. It'd be mashed to pieces. Of course if she is a good rower she might make it."

Mentally I continued to sink.

David broke in brusquely, "Lyn, there's a rowboat. Get in and practice."

Embarrassed by the entire village of Kyuquot now leaning over the rails of the dock listening to our conversation, I didn't argue. I had never rowed before but I was determined not to give the onlookers any more reason to scoff or banter. I maneuvered the oars to see exactly which of my movements with the paddles caused which action, in readiness for the following day's endurance test.

"I wish you'd get that right rowlock fixed; the paddle keeps coming out," I called testily to David when I came close to the dock.

David countered with "What's a rowlock, Lyn? We don't have them in Canada. You mean an oarlock. Your movements will have to be more decisive than that if you don't want to be sucked under the barnacled rocks." He was grinning. That produced another round of ribbing as the watching fishermen chuckled over this unique domestic discussion.

"Spending three weeks on Solander, are they? Why, it's a vertical rugged rock. There's not even a place to pitch a tent." "Even if they got a tent up there's little chance of it staying up in those winds. You get about the worst weather in North America off Cape Cook."

We went to sleep that night with the news that the local Spring Island radar station wouldn't be able to pick up any SOS signal we might be able to transmit on our emergency transistor set.

In the early light of Monday morning we stole down to the docks to board the *Audrey H*. The harbor was like a millpond as we sailed out; the skies were clear and there was a promise of sunshine. Surely we could make it to Solander in this weather. But Wally, without saying a word, put out all his gear for trolling and lowered the stabilizer poles on each side. We waited.

We desperately wanted him to push on for Solander, but you can't argue with a fisherman when he makes up his mind, so we resigned ourselves to the situation. I fumed silently at the delay. David laughed and joked with apparent unconcern as he helped Wally to fish. Within an hour Wally hauled in the long lines to find about half a dozen seven-pound Coho salmon on his hooks. He neatly slit down the underside and threw the guts overboard. They were immediately pounced upon by gulls, which settled on the swell to digest them.

Ten minutes more and we met many of the fish boats coming in. We turned back to Kyuquot. If it was too rough to fish outside, then it was certainly too rough to negotiate Cape Cook.

"We'll try again tomorrow. Meet you at the boat at three A.M.," Wally said as we returned to Kyuquot.

Early Tuesday morning we made our way in pitch darkness to the docks. No sound broke the stillness but the waves from a departing boat slapping against the pilings. It was only five minutes to three.

At 6 A.M. we were still crouching and shivering beside a row of oil drums, waiting and watching the fish boats as one by one they came to life, filled up their water tanks and moved out of the harbor. At half past six the *Audrey H* slipped into the dock to load its two cold and discouraged passengers. David was as impatient as I to get to Solander and I wondered what his comment would be.

"I know somebody who slept in," he joked casually as we clambered aboard with the gear. I was astounded, but Wally nodded and smiled just as casually.

A fishing community on the outer west coast consists of very practical, hardworking, self-reliant people. To be accepted by the community means you have to prove your abilities. From a little previous experience David had known it would be almost impossible to make arrangements for a trip to Solander through correspondence. To arrive in the village would at least be a mark in our favor. Perhaps waiting on the dock for three hours was part of our initiation. To spend three weeks on Solander would mean final acceptance.

We chugged on under clear blue skies in a light southeast breeze west to our destination. In four hours Solander loomed into view. Puffins dotted the water and flew overhead in endless crisscross patterns. The tall green grass covering much of Solander's rocky cliffs was bowing and swaying in gusty winds. The southeaster pounded great waves against the only side of the island where it was possible to land. Leeward the water was calmer, but there was no way of getting up the sheer cliffs with all our gear. David looked at Wally questioningly.

"No, Dave, we must have a westerly wind. You can't land there without an extra man to row the boat. We can anchor at Klashkish Inlet about an hour and a half from here, then mug up at the fish camp in calm water and try again when the wind dies down."

David and I both hated bypassing our island after all the delays, but we both knew the matter was decided. Away from Solander the waves subsided to a gentle swell as we entered narrow Klashkish Inlet. We tied up at a deserted floating cabin and explored its several rooms. In one we laid our sleeping bags and mugged up on boiled eggs, toast, jam, and coffee on the *Audrey H.*

By this time a Norwegian couple had docked on the opposite side of the float and were icing up their fish. Then two chaps in a speedboat roared in from the head of the inlet. They worked at a nearby logging camp and were waiting for the company to deliver the week's mail. Several other fish boats arrived to tie up alongside for the night. While David joined them in a game of poker I ensconced myself between the sleeping bags in the deserted cabin and wrote my diary.

Wednesday morning was clear and our hopes were high. Hoping to get as early a start as was politely possible we packed our gear on the *Audrey H*. Wally called up from the galley: "Would you like a bit to eat before we go? There are boiled eggs ready."

"Well, I'd be more enthusiastic about getting started before the wind blows up," ventured David.

"I know: let me serve the eggs while we go along," I compromised. Wally's face broke out into a charming crinkling of a blue-eyed smile, and we cast off, the radio giving out staccato conversation as fish boats called the *Audrey H* from a radius of fifty miles.

"Solander Isle—*Audrey H*. How's the crew?"

"*Audrey H*. We are having another try at it. Bit of a fog coming in. Wind too southerly to land at the moment."

"Pretty heavy lump down here today. Fog so thick I can't see the bow poles. You'd better take care," came over in a Norwegian accent.

Our eyes strained anxiously for the first glimpse of Solander as the thickening fog reduced visibility to less than a half mile. Puffins appeared more regularly overhead. Surely it must be close now. Even the sea had flattened out, indicating we were on the leeward side of an island. But what island, and where?

David suddenly yelled, "There she is!" A small gray pinnacle became visible for a second, then disappeared.

Thinking he was within a hundred yards of the island

Wally dropped the hook and David readied the dinghy. Wally stopped the motor so the sounds of the surf and the birds could guide David to the island. David shoved off into the fog to explore the shoreline and determine the best point for unloading our hundreds of pounds of gear.

Uncannily close through that curtain of thick fog lay an island palpitating with life. An eerie silence was broken only by the surf crashing on the shoreline and the shrill cries of seabirds overhead.

With the motor stilled we would be able to hear David's calls, but the tide was so strong that Wally was in constant fear we'd run aground on the numerous reefs that swept by at an alarming rate.

I called to David to maintain contact. Three, four, five minutes dragged by. There was no reply. I called again. Still no answer. Wally started the motor and nudged a little closer to the smashing surf line. My eyes strained to pierce a pathway through to the island. Before I had enough time to work myself into a panic a voice behind me called, "Let's get the gear into the dinghy quickly." David had come up on the other side of the boat. I had been looking in the wrong direction.

Gripping the sides of the *Audrey H*, David spoke tersely, "Lyn, you'll have to row me ashore so I can test whether we can climb the cliff to the top from the landing area I think is most suitable."

Uncertainly I lowered myself into the rowing position and prepared to curse the unstrung oarlock. Out it bounced, slewed around, and we had made a circle before I could get the oar back into place. Too soon the rocky bluffs towered above me. The curtain of fog lifted momentarily to reveal air filled with puffins, gulls, guillemots and cormorants, and a barnacle-encrusted channel fifteen feet wide with vertical walls. I tried to remember David's instructions about turning the boat quickly, judging the swells, back rowing one oar and forward rowing the other. Somehow we swelled in on the right angle and David clambered ashore over the barnacles.

I was now alone. The fast-running tide was sweeping me forward onto the barnacle-encrusted reefs. I back-paddled furiously away from a shoreline that could tear a skiff to shreds in a matter of moments. Away from the heavy swells the surface was calm, an area of dead water where the waves prepare themselves for another onslaught on the shore. Then I realized I was paddling too far out into the fog and was likely not only to lose myself but to leave David stranded on an island with the birds. I paddled forward about fourteen feet offshore and lodged myself in a kelp bed to swirl around the long fronds till I heard David's yell.

"O.K., Lyn, now come in straight like you did last time."

Beginner's luck. All I could do this time was come in parallel, scrape past the barnacles, get caught in the down surge, teeter over dangerously and somehow rise above it, then slew the whole boat around the bed of kelp.

By this time David's language regarding my stupidity wasn't fit to print and I was ready to row off and leave him or else dump myself overboard to see if he cared enough to jump in to my rescue. Instead I blamed my tools.

"Why didn't you fix this crazy oarlock? Look, it's jumped out again." By this time I had wriggled it back into place, the boat had swelled onto the barnacled rocks, giving David a chance to jump aboard and start pushing the oars for me. The fifty yards back to the *Audrey H* were spent in tears, quick reconciliation, and mental preparation on my part for much worse to come when we started to unload the gear.

While I tried to keep the clinker from getting sucked under

the fish boat in the swell David and Wally loaded the gear. The lighter boxes were piled behind me as the rower and the heavier boxes were packed in the stern beside David. The plan of landing was to row the boat stern first into the rocks so that David could simultaneously grab a box and jump out onto the shore. Timing was critical and rowing had to be skillful. Correction, for me that meant not skill but luck. Trying not to concentrate too much on the right oar slipping out and praying I'd come in squarely to the barnacled ledge, I rowed in on the next swell. David scrambled out, throwing a box on shore as he landed. Every box had to be carried separately from the dinghy onto the shore and David had to jump back in the boat all on the same wave. I had to judge the wave to avoid being taken into a crunchingly parallel position above the barnacled reef or sucked down a hole with the surge. When the heavier boxes were unloaded I had the extra problem of holding the oars with my legs to reach around behind me and pull the lighter boxes to the stern so David could throw or carry them ashore. On the sixth landing Wally took the oars and I jumped ashore myself.

"Don't forget to pick us up sometime," I called with a silly grin as he hovered offshore.

"I'll come by on the second and see if we can take you off on the third," came his voice from the skiff.

"O.K. I'll collect the birds on the morning of the third and hope we can get off," David shouted as we waved goodbye.

Solander—at last! Or what we could see of it through the fog. It had taken five days of delays but finally we'd landed. Yellow-black boulders rose from the barnacled shoreline. There was a possible campsite on a grassy piece of cliffside about fifty feet above our heads. Puffins cheekily circled and gulls gargled as they hovered around us, watching our cavalcade clambering up and down the rocks. To make a floor for the tent we axed the grass clumps, clawed out the loose stones and built up the crevices. Unable to eradicate a large central mound we pitched the tent by sawing off the support pole to accommodate the height of the mound. We found soil surface just sufficient to anchor the pegs. Our home for the next three weeks was up—though for how long in those high winds remained conjecture. With the axe we terraced out enough steps to be able to walk around the tent and provide a place to store the boxes.

David placed a bucket in front of the tent. "Every time you pass this bucket of seawater and disinfectant, wash your hands," he advised. "To save our drinking water I suggest that whenever possible you do all cooking with seawater."

Before I had even come to grips with the beaks and feet of the other occupants of Solander I had already learned as we crawled up the sharp barnacles that almost every move meant a scratch. And here where every rock and grass blade was covered with guano, infection was an ever-present danger.

"That bucket of disinfectant is probably more important than our food," David warned. "Now let's find a way to the top."

Finding a way to the top was what I was dreading. "Let's go the shortest way," I suggested. "Up there doesn't seem too difficult," and I pointed to the puffins peering over the ledge of the cliff.

"No. I tried that way before. There's only one way and that's around the cliff face, above the bay where we landed."

I don't know what was worse, trying to find crevices for toe and finger holds on the crumbling cliff face or listening to the thunder of the swells as they crashed in at our feet. Hovering between crevices, testing the shale which crumbled at the touch, trying to match David's stride, thinking how many bones I'd break if I fell, and wondering fleetingly why I was here, I climbed slowly and painfully up the ledges where hundreds of cormorants had balanced their mound nests. The fat, hollow-stemmed plants pulled away at a touch, as did most of the rock face, but some security was achieved by hoisting oneself up on the long, wavy grass. From a distance through the clearing fog the hillside looked like a green, grassy golf course. In reality it housed innumerable burrows and tunnels of millions of seabirds. I pushed and pulled in a traverse across the steep hill, often falling into a deep hole—the entrance to a burrow out of sight in the shoulder-high grass.

We had to determine the age of the young puffins in the burrows. The older they were, the stronger they would be for shipping. If we were to find only eggs our expedition was wasted.

"Just stuff your arm in as far as it'll go," David instructed. "If there's an adult in the burrow try to get it to bite your glove —but not the fingers inside it."

I lay down in front of an excreta-moistened burrow and boldly shoved my arm into the hole right up to the shoulder.

"Ouch!" I lurched backward. Beginner's luck again! Hanging on to my fingers through the glove was an indignant and pugnacious puffin, gripping with a massive orange beak as powerful as a vise. Only David was thrilled. My own feelings were numbed by the pain of my first puffin. I obviously hadn't discovered how to pull my fingers backward from the glove tip.

"That's the idea. You can now see what a puffin looks like in closeup. Bring it over here to the sunny slope and I'll take a picture."

As I stumbled through the grass to him the puffin momentarily relinquished his hold on my glove. I used that instant to slip my fingers inside the glove out of range of his beak.

"Hold him against the sun" was David's next instruction. While I was pondering my best profile for the picture, the puffin changed position; this time it clamped down on the fingers of my ungloved hand, not with the straight rapier of the side beak but with the deep cutting pincer of the front tip. Blood spurted out freely as the beak clamped again in a new position.

"Dave, lever his beak up," I screamed. Just as I managed to loosen his beak he raked me with his flashing three-pointed webbed feet. David was torn with commiserating with my tears, wiping up the blood, and filming. I think he did all three. "Great footage," he commented finally. "Red blood looks good in Kodachrome."

By this time the puffin had torn himself loose and dived downward, wings beating busily and orange feet splayed straight out behind. In spite of the pain of my first intimate meeting with a puffin, I could appreciate the striking breeding plumage of this gaudy bird. With vivid orange bill, white cheek patches, and golden tassels it looked more like a parrot or a clown than a seabird. In fact the locals call puffins sea parrots.

Luckily for me, when I hesitantly groped into my second burrow, I brought out a black, wide-eyed, squeaking fluffball of a chick. If any animal fitted into the cliché "bright-eyed and bushy-tailed" it was this puffin. Black, fluffy down feathers sprouted from his slight body, making him seem an animated pin cushion. His bright buttons of eyes blinked in his first look at the sun. David, relieved to see that the chick was old enough to survive shipping, took notes on the stage of his feather growth and then returned him to his dark tunnel.

Puffins nested practically everywhere on the island, wherever there was enough dirt for burrows. By evening we had investigated about fifty occupied burrows and found puffins,

Cassin's auklets, and petrels. The auklets and petrels, delicate soft-feathered black-and-white birds locally called sea swallows, nest in shallower burrows often emerging from the main puffin burrows as side branches.

Fascinated, I asked David how the birds knew whose burrow was whose in this hillside of holes.

That was one of science's unanswered questions, he explained. The petrels and auklets are nocturnal seabirds. They wander the open ocean for most of the year, then in spring they have to find their ancestral home, a tiny rock like Solander, and return to it to breed. Not only do they have to find the island in the fog and darkness but they have to dig through all the waving grass or thick salmonberry to find their own particular burrow out of twenty-five thousand other burrows in the same small area.

"If you're game, let's spend one night in our sleeping bags on top of the island to watch the night flight," David suggested. "The nocturnal birds gather in rafts on the adjacent waters till dark, then they fly to their burrows to feed their young. Come along now, we're almost at the top."

We slogged on upward, passing the crowded gull and pelagic-cormorant cliff whose bald triangular mass rose majestically from the surf zone to reach the summit. Solander's highest peak was crowned with light and circled by squawking gulls that hovered, glided, and swooped in effortless circuits around us. In sharp contrast to the gulls with their easy movements, the puffins busily beat their stubby wings, putting up a masterly effort to cope with the blustery air currents. Their clownlike white-masked faces and golden tassels streaming out behind them in the wind matched their cheeky sideways glances at us as they veered away just out of arm's reach.

The sun set as we descended. David slid down the eighty-degree cliff face on his seat, flattening grass as he went. Not for me. He knew he could stop at the cliff edge; I would probably continue on into the inhospitable sea. I descended more slowly, bumping down on my bottom one burrow at a time.

We added reinforcements to the tent against the freshening wind and I started preparing dinner over the Coleman stove.

"Lyn, where did you pack the dishes and cutlery?" asked David as he groped under the tarpaulin pantry.

I felt sick. "I thought you packed those things," I suggested weakly.

"Come now—you were in charge of kitchen equipment!"

"Well, there'll be less air freight to pay for," I answered flippantly.

At least we were lucky to have one saucepan. We passed it between us—always under a light rain of seabird droppings. We ate out of cans and used our fingers a good deal. Fortunately this was an expedition where weight of food had not been a crucial issue; having been transported by fish boat we could afford some of the canned comforts of civilization.

After our dinner of bread (moldy after our delays in getting to the island), ham, lemon cake, and coffee, I climbed into my warm sleeping bag. David sat at the door of the tent watching the ceaseless flight of birds overhead.

Suddenly he stood up. "My gosh, Lyn, I just saw a peregrine falcon. I wonder if falcons nest on Solander?"

Falcons on Solander might make David's day. What made mine was first getting onto the island and then getting to the summit and back again—alive!

With memories of Barkley Sound summers I was amazed next morning to find the sun shining and the air so warm I could have worn shorts and a sleeveless shirt, except that the falling excreta of birds flying all around us made it advisable to wear a long-sleeved shirt for protection. I was delighted that there was no sign of fog and the air was clear and bright for photography.

So that the puffins would be subject to minimum stress during confinement we intended to collect them only on the last day of our stay. First we wanted to study the birds' behavior, the number of times they brought food to their burrows, what time of day they were the most active, and how long they spent in the burrows. Then we wanted to rehearse our techniques of catching the birds alive and finally make a film of our activities for the New York Zoological Society, which was sponsoring the expedition.

We prepared for an overnight stay on top of the island. David filled and overfilled the packsacks, then set them before me. Looking at the smaller one I asked innocently, "Who is to carry all that?"

"You, of course. You've hitchhiked all over Europe with thirty pounds on your back. That's just the same weight. Let's go."

"I've hitchhiked on a level highway, not a ninety-degree cliff," I protested.

"Well, you'll get used to it. Heave ho."

I hitched on the pack. It wasn't the weight that bothered me but the awkwardness of the poles for our proposed blind and the tripod that were strung on either side of the back pack. I took my first two steps bent over into a crawl, trying to adjust to the unaccustomed balance. Of course if I had known the slightest thing about mountain climbing I would have maintained an upright stance and kept my weight forward instead of bending double and inching my way over the rock.

After running across the cliff face to show me how easy it was, David called out in encouragement, "Look out in front;

your antennae are catching in the cracks. Stand up, so that the poles are clear."

Instead of following David, who had forged on ahead, I tried to find the easiest route for my unique climbing method—on hands and knees. Not trusting the slippery shale, which crumbled at a touch, I tried crabwalking up a gully. Halfway up, the tripod got caught in a crevice and I found myself stuck with hands and feet astride the gully, unable to move up or down. My fingers were straining into tiny crevices and my toes were braced against ridges in the flaking shale.

"Daaaaaavid! I'm falling. I can't go . . ." The rest was drowned in frustrated tears.

David scrambled over to the gully and threw down a climbing rope, which I grabbed. Just then my toe-hold in the shale gave way and I dangled momentarily till David reached down and pulled me up by one arm. My gallant husband says that I described the incident as "falling weightless for an eternity" when really I "only slipped ten feet." All I clearly remember myself is the shock of toes and fingers feeling air, and David's jocular criticisms when it was all over.

"Now you don't want to drop into the water and get all wet, do you? Perhaps you'd rather swim around the cliff and come up the other side of the island? Come on, follow my instructions and you'll get confidence by doing it yourself. I don't want to nursemaid you. You'll never learn like that."

I liked it best where the cliff face was covered by tall grass. For acrophobes like me the height of the grass blotted out the long drop to the sea.

A little later I redeemed myself in David's eyes when I flushed a peregrine falcon from his meal. Searching under the thick grass David emerged with a partially eaten murre, another member of the alcid family of seabirds. "Darn it, we've seen only three murres and the falcon has taken one already. I was hoping to see more murres, although I know they don't nest here. Keep your eyes open for a falcon eyrie anyway."

We crossed over to the summit, then went down the cliffs on the other side until we came upon a sunlit rocky ledge, a landing ground in front of several burrows and perfectly screened for observation. We flattened out holes for ourselves in the grass, had a lunch of nuts and raisins, then settled down for observation and photography. Although rocks crumbled and every plant broke away with a crack, I made myself as secure as possible a few inches from the dropoff, fortunately hidden by the canvas of the blind.

Although the sky seemed black with flying silhouettes it proved extremely difficult to compose a screenful of flying birds in one's lens. We took shot after shot, used up film after film in attempts to capture the busy wings of the plump, high-shouldered puffins circling by and giving us that searching, sideways over-the-wing look. I laughed when, staring at us, they would bump into their neighbors.

One bird in three had a "face full of fish"—silver streamers that hung from both sides of his beak. It seemed incredible that the puffin could dive under water and catch more needlefish while it already had a mouthful of several others. We caught one growling adult to examine the rasping ridges on the roof of its mouth. As the fish accumulate in the bill crossways, they are held between these ridges, with the puffin's tongue pushing up against the top of the mouth to hold them there. A puffin can carry a dozen or more fish, dangling like long cigars from each side of its beak.

Although Solander's famous wind was building to such a crescendo that I had to hold down the canvas while David photographed, the puffins continued to oblige us by landing in front of the camera. As they socialized in front of their burrows, golden tassels streaming behind them in the wind, they looked like a conclave of black-bewigged judges. Except for their deep, hollow growl when you disturb them in their burrows, the puffins are rather silent birds. They jerk, waggle, and point their bills as the chief instrument to convey signals. Long-continued bill rubbing is a ritual to preserve the love bond between mated puffins for the period of their rearing duties. A couple of males nearby nodded affectionately and tried to make advances but the objects of their attentions departed presumably preferring fish to late-season romance. A chick walked to the entrance of its burrow and, stern first, fired its waste outward, nature's way to keep the home clean.

In spring the puffins form great rafts at sea to mate and prepare for the breeding season on their ancestral islands. Once a burrow is possessed, the mated birds, using their beaks as pick-axes and their feet as shovels, dig and deepen their tunnels. In June a single white egg is laid in the terminal chamber at the end of the three- or four-foot burrow on an accumulation of grasses and salal leaves. Both parents incubate the egg for about forty days, but for British Columbia the shifts and incubation period are not exactly known. After the egg is hatched another forty or fifty days are spent in rearing the black, fluffy chick. A young chick grows not into the gaudy summer plumage of the adult bird, but into the plain all-black winter plumage of the parents. When the chick is fully feathered the parents make less frequent visits to the burrow, are less demonstrative, and engage in more leisurely and longer social gatherings in front of their burrows. Finally they desert their young. After about nine days of fasting, the chick emerges from the burrow at dark, flutters down the grassy hillocks, using its breast feathers as a buffer, and plunges into the sea. It will be perhaps a couple of years before it will return to its birthplace.

We tried to film catching an adult puffin in a mist net, which is a fine net strung between two poles into which the bird can fly without being harmed. Despite the gale-force winds which kept blowing the net into knots, we eventually set it up across the grassy slopes. David searched the burrows till he located an adult puffin, which I was supposed to launch on an orbit that would take it straight downhill into the net and at the right camera angle. I did launch the bird but it dived straight into the grass at my feet.

David reminded me tersely that the camera was rushing through the film at sixty-four frames a second and at twenty dollars a minute I was spending a lot of film.

"Why don't I take the camera and you launch the bird?" I countered. "You need the practice," he retorted. But after some argument I exchanged the puffin for the camera and David launched the bird—right into the grass at his feet. He was so surprised that he let the bird continue on its hurtling flight to the sea.

We found a second puffin and the launch was more successful. We got the film sequence—and a big hole in the net. Obviously puffins were too big and chunky to collect in mist nets.

The days passed swiftly and the final three days before our scheduled departure we spent rehearsing trapping and handling procedures. We wanted to know exactly how long it would take to catch enough adults and young, since we didn't want to start collecting any birds until the *Audrey H* was waiting offshore and our return was assured. After timing feeding visits, we found that adults stay in the burrows from four to thirty-seven seconds. This didn't give me much time to trap the adults in the burrows. Scrambling up and down a seventy-degree, two-hundred-foot bluff in even thirty-seven seconds might be exciting but it was hardly likely to be successful. We found that the highest peak of fish deliveries to the burrows was in the early morning and evening.

In the Faroe Islands and Iceland the puffin is a valuable article of food, caught by means of a long pole with a running noose of plaited horsehair. To net the adults from the air David imitated these islanders as well as the early coast Indians but substituted for their noosing pole a net on the end of a mist-net pole. Partly concealed by the tall grass I waved my jacket to stimulate the curiosity of the flying puffins while David stationed himself forty feet downwind.

One in three birds would be so intent on looking over his shoulders at my antics that David could easily snag one out of the air. In one hour he had caught sixteen gaudy adults. As each bird was brought down, David taped its beak tightly to prevent fighting in the crates on our back packs. Six birds to a crate and two crates per trip, up and down the cliffs we clambered, increasing our skill with each trip. After we had the routine perfected we released the trial birds unharmed.

Slipping down through the flattened grass on my seat toward the tent, I yelled to David, "Look down there in the tidal pool near camp. Isn't that a seal?"

On a rock in the middle of an M-shaped pool by our tent, a splotched gray harbor seal pup was sunning himself.

"Get down out of sight," David commanded. "It's a pup of this year. I can't see the mother. Usually at that age the mother stays close. It's possible the seal-pelt hunters have got her. Anyway if we can catch the pup it'll make a great playmate for Sam."

Leaving our packsacks in the grass and trying to stay hidden, we reached the boulders on the tent side of the tidal pools. Hastily David made a rope noose at one end of his mist-net pole. At our approach the pup had slipped into the water, so we each took up a position at the two exits of the tidal pool to the sea.

In a few minutes a head appeared in the middle of the pool and two curious eyes stared at me for several minutes. Then two cavernous nostrils opened wide and closed as the animal silently submerged. David called out from the other end. "If he comes to your end again, try to pull him up by the flippers."

It took much patience to wait for further reappearances. A seal can stay submerged for at least twenty minutes.

"If we get it, how will we feed it? And get it back to Victoria?" I asked, immediately thinking of the practical consequences.

"We've got plenty of canned milk. And there's the holding pen for the birds to keep it in while we are on the island," replied David, who is never at a loss.

"Let's call it Solly after the island," I suggested.

Just then the pup appeared at my feet. I leaned down to its head, almost kissing it as I ventured my hand forward trying to grasp its short front flippers. It slipped away through the water down to David's end. Eventually the pup emerged from a narrow adjacent crevice and swam right into David's noose in the water.

David yelled, "Come here quickly and loosen the rope on its neck. It's caught."

Barnacles forgotten, I clambered over the rocks to free the noose. Solly was fat, pugnacious, and several weeks old. It took considerable effort for David to hoist him out of the water then haul him by the hind flippers up to the tent.

As a temporary measure David put a box over the pup and

then sat on top, exhausted. We took turns sitting on the box and cooking wieners on the Coleman stove. It was so windy that the bread slices blew away as soon as I had buttered them and the coffee saucepan was cold before it had passed between us. No mother seal appeared. Using the food boxes and the holding pen for the birds David rigged up a home for Solly. On our knees against the battery of wind, we threw buckets of sea water over him and tried to make him take down warmed canned milk.

We planned to leave Solander in the next couple of days, so we were not worried when the seal refused to take any of the proffered milk. A seal of his size can go a long time without milk. Back home we could more easily make friends with him. It would probably take a week of force feeding or intubation with a rubber tube before he'd learn to swallow the tube or suck from a bottle. Older than David had thought, Solly would soon be ready to take down whole fish.

The prearranged departure day came and we sat in camp watching for the *Audrey H.* For six days the wind had not dropped below thirty mph and high seas were running. Should we go and start collecting just in case Wally could get into shore? What if he arrived and we weren't ready? Collecting and carrying to the shore would take at least eight hours and Wally certainly couldn't heave to in rough waters for that long. We decided to wait, sure that Wally would not face such seas.

We had to spend the following day inside the tent taking turns to hold it around us in more than fifty-mph gusts. No longer was any part of the tent floor even comparatively straight. All our gear had fallen from the sides on to the central mound where the pole still gallantly supported some shape of a tent, but only while we were holding it. When I wasn't doing tent-pole duty I disinfected my cuts, arranged the films we'd taken, and chatted about our retirement when perhaps we'd find the time to view them. Later I mixed up baking-powder biscuits to cook on the Coleman while David skinned a dead puffin for his scientific collection.

By dusk the wind had moderated. Then the full continuous roar of a boat's motor throbbed upon our consciousness. We peeked outside the tent to catch a glimpse of fishing poles appearing around the rocks. It wasn't Wally, but a fish packer on its way to Vancouver. David scurried down to the water to shout to the captain above the wind.

"Will you phone the *Audrey H* and tell Wally to come to Solander as soon as he can?"

There was no time for more. The packer passed. We hoped the captain understood.

By noon the next day all our gear was stowed. The *Audrey H* was sighted. By signals worked out in advance we let Wally know that all was going well and that we would be ready to leave at ten o'clock next morning. He turned away to wait in nearby Klashkish Inlet.

We started up the cliffs to begin catching adults, only to find that the wind had shifted and the puffins were no longer flying close to the cliffs where we'd been able to catch them with a long-handled pole. We had to try something else.

We tried a systematic search of the burrows. By evening we had only six adults and their young. So much for all our rehearsals and optimism. But if winds shift one way they can shift another, and in the twilight the stream of circuiting puffins again came within netting distance. In minutes David had pulled ten more out of the air.

As the moon followed the last streaks of the setting sun the puffins were mostly on ledges or out to sea. We heard the calls of the night fliers. David had previously dismantled the tent, intending to spend the last night on top of Solander to observe the night flight. This would put us in optimum position for the morning's final capture. With everything packed, we hoped it wouldn't rain. As each grassy hump was the roof of somebody's burrow, there was no room to lay a sleeping bag. Standing or sitting up was essential.

To determine which of the nocturnal seabirds nested on Solander and in what proportions, David strung up a mist net. We felt the whir of delicate dark shapes in the night sky as petrels flitted, swooped, and chittered. From beneath the waving grass we heard the brrr brrr of nesting petrels, probably helping to guide their mates home. The air was so thick with birds that several landed on our heads and shoulders. In a few minutes we checked the mist net to untangle thirteen boreal and eleven fork-tail petrels. The petrel chick is brooded by one of the parents during the day for the first few days after hatching. Then it is left alone although it continues to be fed by the parents at night till it is feathered.

By 4:30 A.M. the sky was lightening, the petrels had departed and the puffins were coming on duty. Burrow after burrow we searched and our collection slowly mounted. The difficulty lay in getting down the cliffs with our loads. Poor David was trying to balance three wooden crates on his back, while all I had to carry was a camera case of chicks and a couple of large empty milk cartons for any more I might find on the way down.

"I don't think you should use that milk carton. The chicks might burrow down and suffocate. Put them down your sweater instead," suggested David.

Easier said than done. The first furry blighter burrowed down behind my pants as I tried to reach into successive burrows. David might be able to carry twenty-four birds on his back and two more in his shirt but I didn't seem capable of keeping even one under my sweater.

As we descended with the last load, the *Audrey H* came into view.

The last of the camp was dismantled and the gear lugged to shore. At the last moment the birds were placed in their separate shipping compartments and the lids were screwed on. Solly was enshrouded in an orange tarpaulin. Then we dashed to the tideline to meet Wally, for now the real race to New York was beginning.

At last I could change out of my Solander clothes that had been blackened by the dirt of the burrows and whitened by the excreta of the birds. For the first time in weeks I removed my cap and combed my matted locks.

The sea was strangely quiet. All was grayish-white—the slight swell, the sky, the white glare of the sun all blended in one monochrome. In four hours we were in Kyuquot, where Lucy and Esco were on the dock to greet us. David sprinted to the airline agent. We had radioed ahead to have a charter plane waiting for us. There were still two hundred miles to go and only four hours to get the birds on a New York-bound jet.

"We couldn't get you a plane today," the agent told him. "Maybe tomorrow evening."

We were stunned. The puffins' chance of survival would be drastically lessened if they had to stay cooped up a whole day. We looked so stricken that the dispatcher took pity on us and rescheduled several charters: within an hour and a half Mike Carr-Harris had a new Canadian Found float plane waiting at the dock.

Birds, equipment, camera gear, and Solly, still in his tarpaulin, were quickly loaded onto the plane. David threw me a toothbrush with the words "Sorry, Lyn. There's no room for you. You'll have to stay here and take the scheduled flight down to Vancouver tomorrow night."

Oh no. Kyuquot may be pretty and full of friendly people but I wanted to see this drama to the end.

"Couldn't I be squeezed in and something inanimate left behind?" I plaintively appealed to the pilot.

He relented. "O.K. Jump in front and David can sit behind with the seal on his lap."

I didn't have to be told twice. In I went in the seat of honor. And out went the sleeping bags and plastic bags of clothes. We crossed our fingers and hoped we would be able to take off with our load. During the trip south to Vancouver, fog and our cramped positions ruled out photography. Whiskers to whiskers, David and Solly had no alternative but to make friends.

We landed on the Fraser River, where a trailer took us up out of the water. It was a mere twenty-five minutes before the Seattle flight when we quickly loaded the boxes from the plane into our waiting station wagon.

We knew all live freight had to be checked in an hour before the scheduled departure, so David had no conventional way to get the birds on the plane. Without a word to me he drove off with his load to the airline terminal and drove directly onto the tarmac to park under the very wing of the jet that was readying itself for the flight.

I remained silent, stoutly holding Solly's hind flipper as he indignantly tried to free himself from the folds of the tarp. A wild seal flopping on my lap in a car parked illegally on the tarmac of an international airport could only make matters worse, I felt—even more so if he flopped through the window onto the runway.

David was determined to cut red tape and get the birds

safely on that plane whatever the cost. Airport personnel were already striding toward us. Luckily the captain got to the car first. He bent his head to the open window just as Solly wriggled free of the tarp and coughed in his face. The captain reeled backward but David had his attention.

"I need your support. I am a biologist and a pilot," he blurted out. "Timing is critical. We've just spent several weeks collecting these live puffins isolated on a barren rock off the west coast of Vancouver Island. We have to get them to New York immediately. Except for an odd pair here and there this species has never been studied or displayed in captivity. They are unable to stand stress. So far it has been an incredibly short time since they were collected. This flight is the only connection to the New York flight out of Seattle. Can you help us?"

Luckily the captain was sympathetic. He spoke quickly to the irate freight officials, who by this time had converged on the car. They peered through the windows at our unique cargo then fell backward a few steps as a seal lunged toward them. Frantically I clutched his hind flippers in one hand. With the other I tried to close the window. David and the captain had run to the freight office. With two arms full of a mad squirming seal I wasn't able to drive the car away: I speculated on spending the night parked on the tarmac of the airport while David was flying first class to Seattle, as it was essential to be sure of the birds' transfer to the New York plane.

In a few moments, much to my relief, David and two of the attendants returned. They started to stow the boxes on the plane. I was overjoyed when David spoke.

"Well, I won't need to go to Seattle after all. The captain is very much interested in coast wildlife. He promised to check the changeover. I'll phone Seattle airport in an hour and a half and get a blow-by-blow description of the transfer."

David climbing to an eagle's nest in a huge cedar

Lyn proudly holding an adult bald eagle she had just captured as it was swimming across the channel

With practice, it's possible to control an eagle with one hand

Proud and defiant bald eagle proclaims his superiority

Facing page:
David bands an adult bald eagle

The eagle's nest the easy way, by helicopter

Toweled immature eagles await banding at study base camp

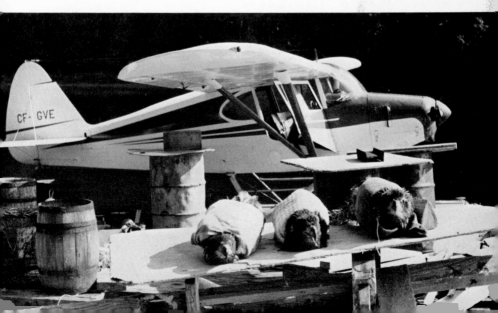

CF-GVE

California sea lions sunning themselves

A mother Steller's sea lion taking her pup to a safe place

"Will I be oiled too?" asks one of the lucky, unoiled, sea lion pups on San Miguel Island

Facing page:
David examines a sea lion pup on San Miguel that died in the oil spill

David overlooking bird bazaars of Triangle Island

David being evicted from the back of a 6,000-pound bull elephant seal

David recording the strange sounds of elephant seals; many of the animals are molting

Murre colony

David with a murre chick he hatched

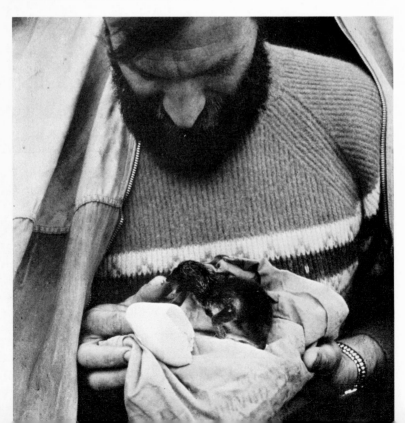

Murres, close up

Salal Joe's cabin and our plane in Barkley Sound

Our good friend Salal Joe hamming it up as he patrols the heavily forested islands of Barkley Sound

One of a pod of killer whales treading water in order to examine the Hancocks' skiff more closely

Islands of Barkley Sound

Our cozy retreat at Dutch Harbor, Barkley Sound

The boxes were loaded and David climbed into the car beside me. By this time Solly was certainly getting the upper hand.

"You look rather smothered with seal. What are you doing on the floor, Lyn? Like it down there?" David laughed, tension over.

"David, do something. My pants have ripped to the point of non-respectability," I complained.

"We'll drive over to the freight office so you can change. You did bring another pair, didn't you? I left my suit in the station wagon so I could go to Seattle," was David's reply.

An amused airlines officer pointed to a smoking room, where I changed into a crumpled but clean pair of jeans. As I looked down at my clodhopper boots I cursed the fact I had left my shoes on the wharf at Kyuquot.

We drove into Vancouver wondering where we would spend the night, since we had rented our own cottage for the summer.

"Let's call at Helen's and sleep on her chesterfield," I suggested, but David decided: "No, we'd better go to our own house before the people have gone to bed. They might let us sleep in the back room for the night. We could then put Solly in Sam's pen."

We drove past our house. It was in darkness. So was Helen's.

"Let's go to the Bayshore," David suggested.

"You sure pick your times! Look at me. Covered in Solander dirt, creased ripped jeans, and heavy hiking boots—and a seal. I've got a better idea. Let's put our sleeping bags down on the floor of your office at the university," I argued.

"Guess where they are? On the wharf at Kyuquot. No, we'll try the hotel. We need a good night's sleep." David grinned. "And you'll give the hotel staff something to talk about, walking across the foyer in those clothes, carrying your toothbrush. I'm going to change into my suit. I'll pretend you're my country cousin."

As it was the height of the tourist season the first ten hotels we tried were booked solid. Finally we had to settle for the Bayshore. With David walking ahead grinning as if my appearance was a huge joke, we took the elevator to our room on the fifth floor. I avoided the eyes of the polite clerks, leaving them to imagine the worst. And I wasn't even wearing my wedding ring! At least I wasn't carrying a fighting seal—we left Solly in the station wagon in the hotel parking lot for the night.

David dived for the phone to check on the Seattle change-over, then to get Mr. Conway, the Director of the Bronx Zoo, out of bed to tell him the puffins were on their way.

When we awoke the next day there was a telegram from Mr. Conway. It said the shipment had come through in fine shape and that the puffins were magnificent. Well they should be: none of the puffins had been in confinement more than thirty-three hours before they arrived in New York, and none of the young more than fourteen hours. Considering the burrow digging, the hill climbs, the unreliable weather, the seven surf landings, the forty-mile bounce on the fish boat, the two hundred-mile charter flight and the Vancouver-Seattle-New York jet trip, we felt pretty proud as we read and reread the telegram.

When I sneaked downstairs to take a look at Solly in the station wagon, I found his face pressed against the glass had attracted a small crowd of sightseers in the parking lot. The bystanders were even more intrigued when I lifted the lid of four boxes in the back of the car and offered strips of herring to four young puffin chicks that David was keeping for his own research.

With me still scruffy in my jeans and hiking boots, we caught the next ferry to Victoria, where the folks were looking after our other animals. Mother and Father were relieved to see us emerge from the car with a mere seal and four puffins.

OUR TECHNIQUES FOR INTUBATING SEALS AND sea lions had much improved since the early days with Sam, when we had improvised with a plastic bag, elastic bands, and a rubber hose. Now we had a specially constructed plexiglass pump that could send up to a gallon of liquid into a seal's stomach at one time.

While David straddled Solly to hold his mouth open, I poked the soft rubber tube down his throat. After making sure he could breathe normally David then gave the signal to pump in the milk. If you're lucky, some seals learn to swallow the tube themselves. To start the weaning process we blended in fish with the canned milk and vitamins. Later we force-fed up to a dozen herring a day. It sometimes takes several weeks before a seal will swallow fish of its own accord. I had become quite used to handling herring but I couldn't face filleting three hundred pounds of goggle-eyed cod and halibut that David brought into the kitchen to vary Solly's diet. Once again I enlisted the help of the neighbors.

Our hopes of finding a playmate for Sam diminished when he ignored Solly completely. A seal has short stubby flippers and propels itself through the water by wiggling its fat body, whereas a sea lion like Sam has much greater maneuverability and swims by using its long streamlined flippers as a human does. Perhaps it was Solly's lack of flexibility on land that disgusted Sam.

I devoted my main energies at this time to the raising of the four puffin chicks. David has always been attracted by the challenge of raising and studying "difficult" animals. Puffins are notoriously "soft" and unable to stand stress. Several zoos had tried to keep puffins but failed. While the New York Zoological Society proceeded to raise their puffins in refrigerated, air- and water-filtered quarters, with an army of trained keepers to watch their every movement, I proceeded to raise my four in the basement.

Each puffin was given a dark rectangular cardboard box as an artificial burrow. I lined the bottom with papers and fine gravel, which were changed several times a day. Cleanliness is the all-important factor in maintaining seabirds. A puffin keeps its own natural burrow clean by backing up to the opening when it defecates, as I had seen.

Five times a day I offered them by hand small strips of herring, cod, and halibut. I then left a few strips in a corner of the box to encourage them to eat by themselves.

Each of the four had a distinctive personality and I named them accordingly. Smart One was the oldest, the most handsome, and the one that caused most of my gashes as she tore with feet and beak into my fingers rather than the strips of fish.

Whenever I approached Panicky One she waddled into the corner in a vain attempt to hide, then furiously flapped her short, stubby wings, pecked and tore at the air, whether there was a fish to grab or not. She ate more fish in fright than to satisfy her appetite.

Teena was the youngest, so tiny and such a ball of down for so long that I began to wonder if her feathers would ever grow.

She had curiously alert and staring eyes, and although she spent most of her time in a huddle, she liked to be handled.

My favorite one was Calm One, not because of any specially endearing qualities but because he was the greatest challenge. He steadfastly refused to eat but calmly sat on my lap ruffling his chest feathers up into a "full-crop position" whether I had force-fed anything into him or not, and practically dared me to keep him alive.

David warned me that if I started force-feeding I would have to continue it always. He suggested I abide by the motto "survival of the fittest." But I was determined to be as stubborn as a puffin. The weeks stretched into months before Calm One would take a fish on his own.

. The most important factor in keeping seabirds alive is to keep their feathers completely clean and waterproofed. After Solander, I would have thought that conditions in the folks' basement were infinitely cleaner than a puffin burrow. To my dismay, when I introduced the quartet to water for the first time they panicked and fluttered to the bottom as if they were nearly drowning. The slightest handling had lessened the birds' natural waterproofing. For the next couple of weeks I alternately allowed them in the water to clean, on land to preen each feather dry, then the whole process was repeated many times.

For the winter session of school and university we returned to Vancouver, where David further taxed our landlord's tolerance by building a puffin tank in the backyard beside Sam's swimming pool. To maintain cleanliness the birds were kept on wire or in the water. The tank was kept constantly overflowing to keep the surface of the water clean of fish, oil, and the birds' excreta. Gradually they acquired waterproofing.

Now David wanted to experiment with diet, so that the

birds either in the second or the third summer of their lives would come into their beautiful breeding plumage. In the wild, puffins eat a variety of fish—sandlance, needlefish, jack herring, and smelt. The frozen herring fed to them in captivity lacks carotin or the red coloring matter that is found in carrots or shrimp carapaces. By feeding a variety of vitamins and carotin we hoped to induce the colorful orange beak and feet.

It is amazing the effect that those four puffins had on me. I remember vividly one Black Friday when I thought I'd lost them. David was away speaking at some symposium when Alan Best, the curator of the local zoo, brought a tiny ancient murrelet, a relative of the puffin.

I remember showing Alan the puffin tank in the backyard and then returning to the kitchen to feed the ancient murrelet. Next morning as I was getting ready to leave for school, I glanced out the window to see the puffin pen doors wide open. I hadn't closed them the previous night. I rushed out to the tank and counted them. Two were missing. With tears streaming down my face, dreading what dogs and cats could do and what David would say, I searched in the grass up and down the lane, then ran inside to phone Alan, who promised to come over with some dogs to help me search.

Next I phoned the principal; realizing that I was in no state to teach that day, he arranged for a substitute. I continued to look through the bushes in the garden until Alan arrived with some volunteers from the zoo.

I glanced inside the puffin pen and there, huddled together in the corner alive and well, were all four puffins! I don't think I've ever felt more of a fool in my life. The puffins must have been there all the time. I apologized to Alan, the other volunteers, and the dogs. All took it well and left.

Sam still refused to have anything to do with Solly, so we

finally found a home for the latter in a large aquarium with more friendly seals for companions.

Not appreciating our efforts to find him playmates, Sam eventually found his own—an orange tomcat called Ginger.

As soon as Ginger arrived, Sam approached with his usual bluff charge. The cat was unimpressed. Sam stopped short. Usually, inquisitive cats, dogs, and people are completely intimidated by Sam's initial challenging charge. Instead, the cat pushed himself up against Sam's whiskered face and purred. A friendship was formed.

We undoubtedly had the most tolerant cat in the whole world. Sam took him by the scruff of the neck and dragged him to the middle of the kitchen floor, then prodded him a few times with his nose as though daring him to run away. If the cat didn't move, Sam shuffled a few feet away and pretended to ignore him, keeping a watch out of a corner of one big brown eye. Cautiously the cat would start to creep away. But, with two quick lunges and a lightning extension of his highly flexible neck, Sam had him back in the middle of the floor. This game would continue for hours at a time.

Sometimes when the placid cat tired of being exuberantly herded around the kitchen by the harem master, he retreated into his box. Then, if Sam couldn't persuade him to come out, he proceeded to push box and cat together around the kitchen. Ginger didn't put up much resistance: he regularly sought out the seal's affection. After a few minutes' solitude, Ginger always came back, purring and rubbing his chin against Sam's neck to start the game all over again.

At mealtimes, both eagerly awaited a dinner of herring. But Sam could wolf down thirty-five herring—eight pounds— before the finicky cat had the first one organized.

Sam treated his cat exactly as a fur seal treats one of his

harem. What made his behavior so interesting is that normally bulls don't become harem masters until they are about ten years old. It is only then that they have gained sufficient fighting experience and the necessary weight, say five to seven hundred pounds, to hold a desirable harem against all comers.

We tape-recorded Sam's mating call, a series of short clicks, and noted that at this time he was less interested in food.

When the cat sought warmth and privacy on the stove with its open oven door, Sam lay next to him, even though the oven was very hot. It made a ludicrous picture to see Sam, wearing the best of fur coats as well as an insulating layer of blubber, sitting on the open oven door basking in the heat.

But the friendship of Ginger and Sam was destined not to last. One day Ginger fell into an oil barrel. When we rescued him he looked so woebegone that we decided an extended holiday for him in Victoria with David's mother might be sad for Sam but a rest for the cat.

6

Mr. Hancock, I Presume?

DAVID'S STUDIES ON THE BALD EAGLE WERE BEGINNING TO attract attention in other parts of the continent. After winning the Howell Award for Ornithology and publishing several scientific papers, he was invited to address the annual meeting of the Canadian Audubon Society, which had sponsored his bald-eagle research. Away he went to Toronto, allegedly for three days; he returned seven weeks later. Taking advantage of every moment, he had accepted invitations to talk his way from Toronto through New York, Florida, and Colorado back to British Columbia.

As well as being in constant demand as a speaker, David was also preparing television programs and writing both scientific and popular magazine articles. Bob Fortune of the CBC Outdoors program *Klahanie*, encouraged him to make his movie camera an integral part of subsequent field trips. Although David had owned a movie camera since he was sixteen, I tease him that he had never had any money to fill it with film until he acquired a working wife.

His television programs and magazine articles now received

such an enthusiastic response from the general public that David realized his own promotion of conservation could be best served through the media of film and print, rather than at the university. It seemed to him that the prime problem facing conservationists today is educating the public to the values of our outdoor heritage, specifically pointing out the meaning and value of nature. People are unconcerned about a species or its environment until they know something about it. David now conducts studies with notebook in one hand and movie camera in the other. I try to tag along as field observer and photo-journalist.

To David the Saanich Peninsula on Vancouver Island is God's Country. All his life he had wanted to live at Island View Beach ten miles from Victoria. Despite the fact that he was still a graduate student and my teacher's salary scarcely covered our living expenses, when the six-acre property on Island View Beach came up for sale he was determined to buy it.

All we had were debts. Our first two debts were unusual for newlyweds: one was a loan for our return air fares from Australia, the second to renew our float plane from the frame outward. The summer before we met, David's plane had pulled loose from its anchor in a gale on the Queen Charlotte Islands and when it was recovered salt water had completely corroded the metal structures.

However by a complex arrangement of loans and finances that baffled me, we found ourselves the "owners" of six acres of flat, treeless swamp situated behind a dike and surrounded by potato fields.

The buildings consisted of two beach cottages, some rambling rooms behind a coffeeshop and restaurant and an ancient Go Kart track. As David signed the papers the former owner commented, "Nobody in their right mind would buy this place." David's parents, saints that they are, let their own house in Victoria and moved in to make a home of Island View until we could leave Vancouver. On weekends and holidays David and I commuted from Vancouver to Island View to knock down the most irretrievable of the buildings and turn the restaurant into a study area and library to house his books, films, tapes, and camera equipment. The living quarters were heated centrally in the literal sense. There was one central source of heat—a propane heater in the living room and nothing at all in the other rooms.

But the beauty of Island View was the view—a two-mile beach at our front door looking across the Gulf and San Juan Islands and on clear days the white ice-cream cone of Mt. Baker in the States.

With the folks, Sam, the puffins, and cat installed at Island View Beach, David picked me up from school on the last day of the term and we drove to the airport to commence our summer's activities.

David is the one who makes all the careful plans and arrangements for each expedition. I am so immersed in school or the animals at home that I usually learn where we are going only on the first day of the trip. This summer our heavy gear was being freighted north via the *Skeena Prince*, one of Northland Navigation Company's coastal freighters. Unencumbered, we were to count eagles from the Piper Pacer along every indentation of the shoreline from Vancouver a thousand miles northward.

Our destination was Klemtu, where David wanted to map out a new eagle-study area encompassing Swindle, Princess Royal, and Aristazabal islands. Leaving rain and clouds behind us at Nanaimo, we should have found the survey flight enjoyable. While David flew his usual cowboy aerial acrobatics around eagle nests, I tried to mark nests on maps and contain my breakfast of raspberries, that seemed determined to make its way into the plastic bags kept at my feet. Oblivious of the rainforests and the jagged mountains of the magnificent coastline I had only one thought in mind—to descend.

We crossed Johnstone Strait to gas up at Sullivan Bay. A gentleman of eighty pulled our wings in to the float while his wife offered me a Listerine mouthwash, hot water, and a dry cracker for my queasy stomach. In all my flying with David along our coast in all kinds of weather that was the only time I was airsick.

Sullivan Bay, a remote isolated float with gas facilities, was a real home in the wilderness. The cabin and guest house were immaculate but the chief mark of distinction was the profusion of flowers growing out of old rowboats and washtubs on the float. It was a veritable garden of roses, strawberries, foxgloves, and raspberries—not that I wanted to face any more raspberries.

From Sullivan Bay we flew north across the myriad islands and inlets of Smith and Rivers inlets to Ocean Falls. Set around the three steep sides of the head of a fjord, Ocean Falls looked uncharacteristically inviting in the setting sun. Much of the year this pulp-mill town is obliterated in smoke, fog, and rain.

In the late light of 9 P.M. we tied up our float plane beside the busy fishing village of Klemtu, our communications base for eagle-study area four. "Place where logs get stuck in channel" is the old Indian word for Klemtu, situated in a sheltered bay on the east side of Swindle Island.

Over the last hundred years tribes from the surrounding areas have gathered here at the "tie-up place" of Klemtu to trade wood, fish, and their services. At one end of Klemtu is the Indian village built up from the beach with boardwalks which function as streets. In these northern coastal communities with dense rainforests at their back door, vehicles are unnecessary. The vegetation grows so quickly that it would be difficult to keep streets and roadways open. Communication from house to house and to the general store, post office, and cannery or down to the dock is by boardwalk. At the other end of Klemtu is the crumbling salmon cannery of J. H. Todd and Sons. As in most communities where the infinite sea is thought to be the cure-all for all man's wastes, unfortunately the tidal flats at everybody's front door are the garbage dump for the village.

The cannery manager, typically hospitable to strangers, invited us to stay the night in "Nob Hill"—a row of rooms used by the cannery and office workers. Our boat, motors, and study equipment had been unloaded from the *Skeena Prince* the night before and were waiting for us on the dock.

Now that the cannery had closed for the day the action center of Klemtu was down at the docks with the fish boats and the float planes. As David chatted to each of the friendly fishermen about their catches and the local wildlife, I learned of our plans for the summer.

David hoped to track down the rare white Kermode bear on Princess Royal Island and if possible gain some information on their limited numbers. Both eagle and bear studies would be filmed for television. Even if we didn't see the white bear, Bob Fortune from CBC said a program could be prepared from interviews with the Indians about bear stories and the trials and tribulations we might have in tracing their habitat.

We continued along the dock from fish boat to fish boat. As we were passing the large fish packer the *Wamega*, a friendly voice called out over the side, "Brought your pet seal along?"

That was our introduction to Anthony Carter, skipper of a fish packer for three months in the fishing season and writer-photographer for the rest of the year. For the past three summers he had been living in the Klemtu area researching the history and legends of the Kynoc and Kit-is-stu Indian tribes. In his quiet and unobtrusive manner he'd gained the confidence of the natives as few anthropologists have succeeded in doing. Apart from packing fish and learning legends, a day's work for Anthony could well include rescuing passengers from a sunken or burned aircraft or a capsized gill netter, transporting people or equipment, offering a pair of warm socks or a dry bunk to a drenched fisherman or eagle watcher or pouring powerhouse coffee from an ever-ready pot.

My favorite picture of Anthony shows him sitting before the steering wheel of the *Wamega* and playing his fiddle, while the sixty-foot boat plows through a seven-knot current that most sailors would avoid, in a narrow channel you could almost touch on either side.

We talked and laughed with Anthony over coffee in the warm galley until the early hours of the morning. He parted from us with the words we were so often to hear from him: "You just call on me if I can do anything to help."

It was rather disconcerting to follow David next morning into the cookhouse under the gaze of fifty pairs of eyes— Indian, Chinese, and white. I may have been the only female in the cookhouse but I dived like a man into huge piles of hot cakes, bacon, and fried eggs. I thoroughly enjoyed myself even if I did put soy sauce on the hotcakes instead of syrup, under the unabashed stares of the cannery workers.

We needed to find a camp for the summer to store our equipment and to have a roof over our heads to get dry after a day's eagle studies and a safe anchorage for our skiff and float plane.

After breakfast we flew down Tolmie Channel and Kent Inlet to investigate a few of the cannery suggestions, but at all places docking the plane was the main problem. Then about thirty minutes by plane from Klemtu we circled over a float cabin at the entrance of well-protected Qua Qua Inlet in Kitasu Bay. A family of seven Indians looked up from their nets and boats as we taxied in to tie up to the float. It was a perfect place for the boat and plane, but sharing a ten-by-ten shack with seven other people didn't seem too likely.

We were warmly welcomed by Pat and Lucy Brown, their children Patsy and Pat Junior, brother Sam Brown, and cousin Chester Green. David explained his interest in eagles.

"Oh, yes, eagles are protected by law, aren't they?" Pat commented casually.

David looked surprised and pleased to hear that the family was informed, but his expression changed abruptly at Pat's second remark.

"I shot three of them yesterday. You can see two of them from here. One is on a log and one on the beach. An Indian can shoot anything on his reserve, you know."

Expecting David to explode into a round of criticism, I was amazed to see him adopt an attitude of studied unconcern. Obviously he thought this was the best way to handle the situation.

"I am interested in Kermode bears too," continued David, hastily changing the subject. He was further jolted by Pat's next statement.

"My father caught two white Kermode bear cubs many years ago, about 1937, I think. He lost one but the other was bought by an American tourist who took it to Seattle. It finished up in Victoria eventually, I think."

As a boy David used to visit Old Kermode in Beacon Hill Park in Victoria. He was able to tell Pat the rest of the story.

"The Museum and Parks people in Victoria felt that the bear, the only white one known to exist, should be bought back from Seattle. It was raised and kept in Victoria at Beacon Hill Park for many years."

When Lucy invited us into the tiny cabin for coffee I was astounded to find we were to sit down to a real lunch of salad, cheese, meat, sandwiches, and cookies. The Browns seemed keen to join us on some of our trips and invited us to share their float with them. Lucy even insisted we sleep in their double bunk while she and Pat slept outside in their little eighteen-foot cabin cruiser.

The Brown family enjoyed unique hereditary drag seine fishing rights at the mouth of the Qua Qua River. Ordinarily white fishermen are never allowed to fish right into the mouth of a salmon river when the salmon are ready to spawn. Boundary markers are usually posted at some distance from the mouth of a river to allow the fish to mill around without constantly running into nets.

However, according to the Browns, Queen Victoria had passed on to their grandfather the rights to seine up into the mouth of Qua Qua for sockeye, the most valuable of all five salmon species. The Browns were further favored by the fact that the canneries are always trying to woo those fishermen who are skillful at catching sockeye salmon. J. H. Todd and Sons ran both the cannery at their own village of Klemtu and the store at which they bought all their year's supplies. Subsequently the store gave them very lenient and extended credit and the cannery supplied a boat if they were in trouble, picked up their fish almost daily, and saw that the family reached their cabin a few hours before the beginning of the very short weekly season.

The fishing season lasted from six o'clock on Sunday night till Thursday night at six. As of Thursday night the Browns would pack up and head out in their little cruiser forty water miles to Klemtu for a rousing weekend of partying.

Our accommodation problems solved, we flew back to the cannery and loaded our gear onto the *Wamega*. The sun shone brilliantly during a relaxing three-hour cruise along Tolmie and Meyer's Pass. Alfred Hall, Anthony Carter's deckhand and formerly a seine-boat skipper, took the helm while Anthony helped us to film. Alfred was a kind man, hereditary chief of the Kit-is-stu tribe. He told us that the native name for Meyer's Pass was Porbouquas or Home of the Hairy One, that huge hairy manlike ape or apelike man that has been reported for years along the coast from Alaska to Mexico. In California they call it the Big Foot. In British Columbia the Sasquatch. Half jokingly, we scanned the steep forested slopes of the pass for Sasquatch.

Numerous cups of coffee and many eagles but no Sasquatches later, we arrived at Qua Qua to be again greeted warmly by the Browns, who had just hauled in the line with a hundred sockeye salmon. The gear unloaded and the Brown catch packed aboard the *Wamega*, we returned to Klemtu for the float plane.

After other summers of living under cramped canvas, in mossy shelters at the base of eagle-nest trees, in sleeping bags on bare windswept rocks, Lucy's double bunk inside a crowded but cozy cabin was extremely civilized. She spurned our instant coffee, our dehydrated and canned foods, and offered us instead brewed coffee, steaks, canned shrimp, freshly caught salmon, and desserts.

In the evening after our daily boat or plane surveys Pat and Sam regaled us with stories about eagles and bears and Sasquatches. Sometimes we'd hear the same story again and again, but we enjoyed it all the more with each telling.

When the sea-fog hung wetly around like an unwanted blanket and the visibility was nil we postponed eagle and bear surveys to jig cod to use as bait for our eagle traps or boarded one of the fish packers to give assistance and learn their trade. Life aboard a fish packer is relatively simple. The packers travel a hundred miles or so from the cannery each day to search out the separate independent seine boats or gill netters. The cannery packers are tenders that save the valuable time fishermen would otherwise spend running between the cannery and their favorite fishing spots.

We came up alongside the *Irene May II*, which, with the *Wamega*, services the J. H. Todd cannery at Klemtu. After the usual exchange of greetings and comments that today's catch was not as big as it should have been and certainly not like the good old days, the men got to work. Each species of salmon, each commanding a different price, had to be kept and weighed separately.

First the fish were thrown from the hold of the fish boat onto the deck. From there they were pitched with a long sharp pole across to the weighing scale of the packer. Since I had no desire to stand around knee-deep in fish on the boat deck I was given the job of standing back with a small counter to keep track of the exact number of fish that were weighed on each scale. A small gill netter might have caught five hundred pounds of sockeye, seventy-five pounds of springs, twenty-five pounds of dog salmon, fifty pounds of Coho and perhaps two hundred pounds of pinks or chums.

Some days we helped the big seiners. Their daily catches were weighed not in hundreds of pounds but in terms of ten or twenty tons of fish, which had to be pitched across the deck and weighed. Naturally it took few seine boats to fill the hold of the packer. Sometimes with the hold filled to capacity the fish were piled up on deck level to the gunwales, leaving little freeboard and giving the boat an appearance of sinking. On days like these I would be found in the galley.

On the other hand, David, filled with ambition, insisted on skimming out into the bay in the aluminum dinghy to haul in a thirty-pounder himself. Up and down we trolled with Sam, removing kelp and disentangling our lines when they crossed as David delightedly experimented with various lures and weights.

When trolling proved unsuccessful, David drifted beside a reef to jig for cod. With Sam aboard to help, I consented to throw my line overboard, praying as usual that there would be no answer to my jigging of the line. Luckily for me rock fish are normally placid and lethargic and rise as if already dead. Despite efforts to the contrary I caught my first fish—a ten-pound ling cod. Steeling myself I got it aboard then closed my eyes while Sam gaffed it, removed the hook, and handed me back the line to try again. Four three-pound black bass rose to my unwilling line before David decided to return to the cabin. I threw the bass to the eagles perched along the shoreline. David hoped to accustom the eagles to picking up food we offered, for later photographic opportunities.

Next to eagles, David's second love was alcids, the family of diving seabirds to which the more familiar puffin and murre belong. On the way back to Qua Qua David was surprised to see so many marbled murrelets with their chicks. Like all ornithologists he was always hoping to find the elusive nest of the marbled murrelet, the only bird in North America whose nest has not been specifically located. Facts about marbled murrelets are few. The adults drift about in rafts or large groups on the sea as night falls. Finally they flutter off to their nests over wooded country when it is too dark to trace their flight accurately.

A close relative of the marbled murrelet is the ancient murrelet, another alcid about which very little is known. Like the auklets they mass in rafts offshore, then hurtle through the dark to land with a crash in the underbrush and scurry off to their burrows. What I find interesting about the ancient murrelets is that the night after hatching the chicks leave the nest as mere balls of down the size of a man's thumb and scramble down to the sea under the cover of darkness in response to the cries of the adults awaiting them in the surf.

David wanted to experiment with the diet of these smaller members of the alcid family to see if they could be adapted to eating fish in addition to their usual shrimp and plankton. The bird research department of the New York Zoological Society was planning a large-scale investigation into the aviculture, behavior, and breeding patterns of this fascinating group of birds.

With the abundance of easily acquired fish in the area, David decided to catch a couple of ancient murrelets. When chased on the water some birds will dive, some turn around to attack the boat, and some will fly. We were lucky. David revved the motor, two young ancient murrelets skimmed the surface of the water in flight and Sam was just able to scoop them up in the dip net. I installed them in a small dark box and we headed for the nearest troller.

The skipper expressed surprise when we asked him for the stomach contents of a salmon (although I think the fishermen around Klemtu had by now all heard about "that crazy couple of birdwatchers"). From the salmon we retrieved a couple of semidigested smelt and some roe.

Like a madcap cowboy of the sea David spent the next half hour chasing a school of herring as it balled to the surface. Thump thump thump, whacked the aluminum bow as with my feet splayed sideways I tried to ride in rhythm with the bouncing boat. Spray fanned above us in a circle. I wondered casually how we'd support the camera equipment if we capsized. Shaking like a leaf, Sam somehow managed to scoop up a full net of herring as David zeroed into the ball of surfacing fish. He was noticeably quiet when we returned at last to Qua Qua.

Back at the float the Browns were intrigued by the two ancient murrelets. Despite the current emphasis on conservation many people still think that wildlife moving is wildlife to be shot. David's theory is that bringing animals close to people will develop a desire to protect them.

Placing the tiny black-and-white birds carefully in a little water at the bottom of an unused barge, David explained how easily seabirds lose their waterproofing by handling or oiling. He showed the children how well the birds were adapted to a life on the sea, how the position of their legs, so far back on their bodies, made walking difficult, how they swam with their wings under the water.

We offered them the semidigested smelt and herring but they brushed it aside. While Lucy and I prepared another salmon supper inside the cabin David patiently sat in the barge and offered them tiny strips of roe and salmon. After supper he tried again. Still the murrelets refused to eat. Despondent, he idly dropped a strip of semi-digested smelt at the feet of one of the murrelets, and unexpectedly the bird suddenly bent down and gobbled it up. Elated, David danced up and down the dock, the light of discovery on his face. "Ancient murrelets don't have to be raised on plankton. They accept fish as well. Now if they thrive on it we'll be able to study them in captivity."

The Browns smiled too, more at David's antics than at his discoveries. Pat Junior and Patsy were proud to keep a careful eye on the murrelets and offer them salmon and herring strips while we were away during the day on surveys.

As soon as the weather permitted we prepared for our first full-scale aerial survey of the study area. Maps were readied, pencils sharpened, the plane gassed up, and ropes untied. David pushed the starter button. There was no answering roar: the battery seemed dead. Although it's an action that sends shivers up my spine he flicked the prop by hand several times till it whirred into action. The sudden blast of air surprised poor Patsy as she staggered back along the float into her mother's arms.

The engine spluttered to a halt. David removed the seats to examine the battery. The heat of the last few days had lowered the water level in the battery. After filling it with fresh, unpolluted creek water he flicked the propeller again but this time there was no life. It was such a good day for surveying that we didn't want to waste it in plane repairs so we altered our plans in favor of the boat. After bailing the boat, attaching the motor, and reloading our gear, David tried just one final flick of the plane's propeller; unexpectedly it came to life and stayed alive.

For the next three hours we surveyed the shoreline of Price, Swindle, and Princess Royal islands. We skimmed the treetops, scanning the beach and trees for eagles and nests. David's eyes were doing double duty, hoping to see a white

Kermode bear on each river delta, logging slash, or open rock cover.

Exquisite Laredo Inlet cuts a long finger through Princess Royal Island. Once past Quigley Creek we entered the main twenty-mile stretch of the inlet. Mile after mile, eagle after eagle, we flew between towering snow-covered peaks and glacial streams that lined each side of the steep fjord, until at the very end of the inlet the mountains came together to drop directly to the sea. Arnaup Creek scarcely had room to wiggle its way up to a patchwork of little lakes that cringed between the towering rock bluffs. Indians at Klemtu had seen Kermode bears on this creek and we planned to return later on foot.

Working now southward down the west side of Laredo Inlet, David turned the plane into the Bay of Plenty, a small two-mile channel off the main inlet that drains a considerable section of interior country. A Kermode bear had been shot here by an American hunter during the previous fall.

Once we were out of the inlet, fish boats again dotted the sea beneath us. Laredo had been designated as a fisheries reserve to keep it an unexploited safeguard, in case present fisheries management practices were not successful.

Aristazabal Island was different from most of the other islands we surveyed. Just inside the coastal fringe of trees which support the eagle nests, the island opens up into flat, swampy peat moss bogs like tundra, dotted with lakes and lagoons. Deer abound on the island.

Seaward to the Moore Islands we counted many pairs of bald eagles. All but two of the nests were empty. Had the eagles produced and lost their young or not produced any young at all? It seemed most likely that they had been un-successful.

All the other nests in the Klemtu area contained young that

were only half or two-thirds grown. Yet, on our arrival in early August, there had already been many fledged eagles of the present year on the wing. To David this could only mean one thing: since all the young Klemtu eagles were still in the nest, the flying birds must have come from further south.

David's studies on the eagles in the southern part of coastal British Columbia had shown that the young eaglets are fledged from the nest and leave their breeding territories with the parents by mid-July. The exodus is so great that an area which may have produced two hundred eagles in mid-June might have only one eagle in mid-July. It appeared likely that the thousands of birds in the whole southern region had flown north. Were they taking advantage of the salmon spawning runs, which didn't occur in the south until later in the season?

Our fourth study area centered around Klemtu included an extremely varied habitat structure for comparative study to determine what factors initially attract eagles to a particular site. In a few hours we had flown along steep-shored inlets which penetrated the coastal mountain ranges, salmon spawning streams that flowed into these inlets, sheltered inner islands, and exposed windswept outer islands.

David wanted also to evaluate the hypothesis that as human activities and disturbances increase, the eagles respond by placing their nests higher up trees and in better concealed locations, concealed at least from close ground observation.

He had found that some human activities, on the other hand, had contributed to attracting eagles. Near the cannery David observed nesting pairs which regularly receive food supplied by fishermen. Although the small sample size was not statistically significant, our preliminary data suggested nesting pairs that heavily utilized food supplied by man tended to show increased reproductive success.

David hoped that Klemtu, where the eagles were more abundant and expendable than their southern counterparts as study material, would be a suitable area to test certain activities disturbing to eagles, such as climbing to nests, banding the young, and surveying by plane and helicopter.

Unlike the southern study areas, which were closer to large centers of human population, we expected the more isolated areas of Klemtu to be a pristine and untouched control area. We hadn't spent long at Qua Qua before we realized that many fishermen, Indians, and hunters were notoriously detrimental to the local eagle population.

Idle time was often a problem for the younger fishermen. While they were sheltering at anchor in bad weather or sitting waiting for their nets to catch fish, they often relieved their boredom by taking potshots at eagles. Unfortunately a living target always seems more desirable for practice than a floating can. And the abundant eagles with their distinctive white heads and large, bulky nests make prime targets.

The first eagles shot at Qua Qua were only the beginning of a steady stream of needless deaths that we encountered. On one occasion while traveling in the boat we found an eagle floating belly up in the water. On closer examination we saw that it had been shot by a high-powered rifle. The eagle was a nestling, not yet fully grown. I was horrified at the possibility that the eaglet had been shot in the nest.

David followed back the general direction the bird had probably drifted to a nest on the shoreline of Swindle Island. After about an hour and a half of laborious climbing David found the site of destruction. Some of the branches had been shattered by rifle fire. Inside the nest was another dead eaglet. Quite likely the eagle we had found on the water had been shot on the edge of the nest and had dropped to the sea below.

Two other nests we discovered later in the season also contained shot young. We wondered how many more deaths by shooting had occurred in this vast wilderness. The eagles' vulnerability to predation by man vastly increased as the birds congregated on the salmon river to feed on the spawned-out salmon. We were to visit such an area of ravage and destruction later on in the summer.

After locating 268 nests in our few hours of flying we returned to Qua Qua. Sam proudly announced that he was going to take us to the one eagle nest he had observed in the vicinity.

The setting for this nest was unique. Sam led us behind camp up the inlet to a picturesque waterfall at the entrance of a salt-water lake. As the water cascades directly into the sea, the falls increase in height when the tide drops. In winter, when the tides are at their highest, the salt water from the sea actually runs backward into the lake. The lighter fresh water stays at the top of the lake and the heavier salt water drops to the bottom of the lake, making life complicated for creatures living there.

Eagles circled overhead, Steller jays called from the woods, kingfishers hovered beside us. Even a hair seal curiously nosed along in front of our skiff. After portaging the boat, the motor and the gear over the logs at the top of the falls, we headed to some picturesque little islands in the middle of the lake.

Mew or short-billed gulls nested here. David noticed a young mew gull paddling frantically in the open water well away from the protection of brush and logs at the edge of the lake. An eagle hovered overhead. The parent gulls were diving at the eagle in an effort to distract the predator and protect their young. Although the role of the predator is to eliminate the creatures that act abnormally, we decided to interfere. The gulls appreciated our efforts even less than the eagle. One parent tried to distract us by a broken-wing act. Sam was given the honor of netting the spunky, mottled puffball from the water and releasing it in the shelter of the driftwood beside the shoreline.

A few miles further down the lake we arrived at Sam's eagle nest to find that it wasn't an eagle's nest at all. It was much more unusual, an osprey's nest. Eagles place their nests on the side limbs at the top of the tree while ospreys choose the very summit. The nest was unoccupied; perhaps the owners were dead.

Next day we packed our supplies for a week's intensive investigation of Laredo Inlet by boat and foot. After being painfully sunburned at Qua Qua the day before, soaked to the skin while crossing windy, wet Laredo Sound, and exhausted through hauling our boat across the shallow flats of the Bay of Plenty at low tide, we thankfully reached our intended campsite at Pyne Creek.

In this area a year ago, a game guide had led an American hunter upstream for a day's fishing at a lake. In addition to his fishing rod the hunter carried a rifle. As they approached a large beaver dam beside a clearing in the trees, the guide saw, no more than seventy-five feet ahead, a huge white male Kermode bear. He quickly signaled to the hunter, who with one easy shot stepped forward to take for himself the only specimen of Kermode bear ever collected under a hunting permit. The animal had been skinned and the skull collected but the bones were left by the water's edge. David wanted to see if there were any remains that could be of use to science. We both wanted to see and photograph a whole Kermode bear.

Slowly we started up the creek. I stumbled first on a dead

eagle, then shortly up the trail two more rotted eagle carcasses. Later a fourth and fifth, followed by two bear carcasses. With mounting horror and revulsion we pondered the reasons for this scene of destruction. What gun had taken such grisly toll of twenty-one eagles and seven black Kermode bears? Most were rotted. Some lacked skulls. All had their hides still intact. There was reason perhaps for the bear slaughter, but what motivation for so many eagles?

Several months later David learned from Bob Titmus, a local fisheries guardian and Sasquatch hunter, that a Federal Fisheries Officer had been responsible for the slaughter. One of the old school, he had considered that any creature eating a salmon was an enemy. Such killing was inexcusable, and when David reported it later to the Fisheries Department, the officials were most apologetic and quickly pointed out that it was no longer the policy of their branch to shoot every eagle and every bear on a salmon river. It was obvious that not all the fisheries inspectors were convinced of the more enlightened approach to game management now in force. It was some small consolation to learn that the officer responsible for the debacle was shortly after dismissed.

With very few fresh bear signs evident in the area and feeling nauseated by the rotted carcasses we headed back out the Bay of Plenty to the head of Laredo Inlet, where we had been told of a recent sighting by an Indian of Klemtu. Earlier in the spring, while hunting deer, the Indian had seen a white Kermode bear foraging the grass meadows at the mouth of Arnaup Creek. At his arrival the animal had quickly disappeared.

We pitched our tent on a grassy ledge beside a tumbling waterfall. Snow-capped peaks rising on all sides made our camp idyllic—that is, until the mosquitoes and horseflies discovered us. They were more than a match for our insect repellent; within an hour seventy-five mosquito bites covered all the exposed area of my face and neck. I mixed up a sticky concoction of baking soda and vinegar which I faithfully pasted on every bite. David with the protection of his beard commented that he was thankful we were in the wilderness in case anybody who saw me thought I had contracted some dread disease. He confessed, "Kissing you is like taking a dive into a barrel of pickles."

We temporarily escaped the brunt of the stinging horde by driving the boat up to the face of the falls to fill a couple of saucepans with fresh water. I then conducted cooking operations from inside the tent, reaching out through the flap to tend the Coleman stove.

Next day the poetic inspiration of the setting was hampered not only by the ever-present mosquitoes but by the harsh realities of humping a week's supplies through the brush to the top of the falls.

Negotiating waterfalls in this steep country can be difficult. The Indians of Klemtu had described an old trail that led through the tangle of the forest edge to level land above the falls. Our instructions were to look for a marker about halfway between the ocean and where the falls started. Then a quarter of a mile inland we were to look for a fallen log marked by an X. At the base of this log was another tree with a blaze mark and the trail followed clearly on from there.

Laboriously we toiled upward through the thick brush. Following behind David with my eyes more on my shoelaces than on the scenery, I was at first oblivious of our arrival at the falls. When I looked up I reached immediately for my camera.

What a spectacular spot to film a Kermode bear if one came clomping across the log that bridged the top of the rushing waterfall!

"O.K., let's go, Lyn. I guess the trail won't take us above the waterfall without crossing that log. Why don't I go over first and then film your crossing?" suggested David blithely.

Vainly I looked for some alternative route while David, expertly balancing both his gear and some of mine, tripped lightheartedly across the log to the other side. Now if the log had been lying on a sandy beach and not slung across a hundred feet of thundering water I could have strolled over it in easy abandon for any camera. I started on the perpendicular, but was soon down on all fours peering like an abject coward at the yawning abyss of rushing spray beneath. David yelled encouragingly, "Hurry up and get over here. You're wasting the film."

Crouched in the middle of the log, and unable to go forward or back, I doubt if even a Kermode bear could have moved me. At such times David's psychology is to pretend to go on his way, on the theory that I will probably follow if I am abandoned. Desperately imagining that the log was laid out at home on my living-room floor, I changed my position to sit down and straddle the log. Eventually I reached the other side by thumping along on my backside. I like to keep my center of gravity close to terra firma—or logs.

As it happened David hadn't deserted me. He had stayed to capture my ignominious traverse on film.

Above the falls, the densely forested mountains loomed precipitously but the land beside the lake there leveled out to make hiking easier. I was intrigued by the activities of the industrious beavers which had freshly shredded the alder and birch along the shoreline. Trees about six inches in diameter were neatly felled with typical beaverlike precision. They had chewed into the tree at an angle of almost forty-five degrees from both the top and the bottom so that when the tree fell it butted together to make an angle of ninety degrees.

No beavers were apparent in the daylight but we did flush a brood of blue grouse chicks, which with their very tame mother were strutting up and down within a few feet of us. They showed only a mild concern until David whipped out a length of nylon noose that he always carried in his pocket, hooked it to the end of an alder branch and snared one of the unwary chicks.

After taking some closeups I put the chick in my sweater. Its occasional peeps caused the female and the rest of the brood to stay close to the camera. At one time in response to her chick's peep the mother flew around us flapping her wings as decoy. The film sequence finished, I released the chick and the family speedily disappeared into the wilderness.

Night came suddenly in this steep valley. The aspens rustling in the wind echoed the crackling flames of our campfire. At midnight a wolf howled. A few minutes later other wolves took up the chorus. We pulled our supplies beside the tent. The calls came closer but no animals could be seen in the stark blackness of the night. In the morning when we awoke, I was awestruck by wolf tracks within the guylines of our tent. They had wandered like ghosts a foot from our sleeping bags yet had made no sound and caused no problem.

After several days of humping our gear through a jungle of dense undergrowth, swinging on cedar boughs, stumbling through salal and huckleberry bushes, investigating bear and deer tracks, fishing for food in mountain streams, David finally decided that Kermode bear sightings might be more fruitful from the plane during our eagle surveys and we returned to Qua Qua.

Remembering the television programs, we took along the movie camera on our next survey flight. Filming from a float plane is not easy, especially when eagle observations have to be made at the same time.

Trying to film through an open window or partly opened door of a float plane with the wind rushing in my face and ears, leaning far back in the seat to compose a foreground of the plane's instrument panel and struts, changing film, exposures, and lenses, rewinding every seventeen feet, and then remembering the eagle observations called out by David to be later recorded after the film sequence was finished, were all too hectic for efficiency.

David interrupted my frantic movements with "Let me take the camera. You fly the plane."

But not for long. In theory steering a plane in a wide expanse of atmosphere seemed less hazardous than steering a car on crowded streets. In practice the treetops came rushing up toward me at an alarming rate and I soon handed back the controls to the pilot.

"Let me take the camera *and* the plane" was David's final comment, and for a time he tried maneuvering the controls for scenic shots with one hand while the other one was stretched behind his head to hold a camera suspended between the fingers which balanced it and those which pressed the trigger. I hoped the lens saw what the eyes were not in a position to see. At one particularly hazardous moment we found ourselves both changing film—David with the movie, I with the stills, and the plane left to its own devices.

When chill winds whipped up the whitecaps to limit our

activities to the beaches, we snared adult eagles for banding, measuring, and release. The heavy rain caused new rivers to find their way to the sea; the waterfalls behind camp rushed madly over the newly exposed rocks at each low tide. The Browns gave up fishing in such downpours and invited us to a salmon barbecue on the beach. We were glad of an opportunity to film an authentic activity in the daily life of a Kit-is-stu.

Sam as head cook cut off the head of the salmon, took out the backbone and splayed the fish on thin cedar sticks over a fire of beach driftwood. To achieve excellence he cooked it slowly, repeatedly turning it to let the juices run back and forth across the flesh to keep it moist and succulent.

To while away the time, Pat collected the long palmlike fronds of the devil's club, a common plant in the coast forest. Hikers find to their dismay that it contains many sharp, poisonous barbs. Pat told us that for the Indian the devil's club has many uses. He took one stem and cut it into four parts (four is the mystical number of the coast native tribes), then placed a foot-long piece in each corner of his boat to bring him better luck during fishing. Pat claimed that a piece of devil's club carried in the hand during hunting would bring good luck to the hunter. When the bark of the plant is dried, ground, and brewed into a tea it is believed to give relief to such ailments as the common cold, TB, and syphilis.

While we were waiting for supper a large seine boat appeared in the bay in front of the beach. A crewman in a small boat headed to shore to snub a line against the trees. The seiner chugged away from the shoreline to set the net in a semicircle for the legal twenty minutes. In the meantime the crewman on shore had been beachcombing. On return to his boat he found it had drifted out to sea and came to enlist our help. We re-

trieved their small skiff for them and watched the big seiner
purse up the net and bring in their catch—five times more jelly-
fish than salmon. While we were commiserating with them the
Browns were trying to attract our attention by waving and
shouting on shore.

The barbecue was ready to be served, and Lucy surprised us
all by suggesting we take the salmon home to the cabin and eat
it properly with plates, cutlery, and condiments. But after the
long wait and the delicious smells of the salmon cooking this
postponement seemed out of the question, so Sam proudly
served the salmon on plates of freshly picked vanilla leaf. For
David and me it was the most perfectly golden succulent salmon
we had ever eaten. The only drawback was that we felt at that
point we could easily have eaten a whole salmon apiece.

The next morning we were scheduled to leave Qua Qua for
Klemtu to meet the Federal Fisheries research vessel, the
Howay, which was to take us to Triangle Island, the largest seabird rookery in the entire northeastern Pacific, where we were
to study, photograph, and collect murres for the New York
Zoological Society. Earlier, in the spring, David had spent a
month on Triangle Island observing and filming the breeding
biology of British Columbia's only ground-nesting bald eagles.

As we traveled through Meyer's Pass on the way to Klemtu
a sleek gray vessel 115 feet long drew alongside. It was the
Howay herself. We were amazed to see such a boat come
through this extremely narrow channel. Although we waved
excitedly to the officers standing on the bridge and they waved
back, David wasn't sure they realized that we were the ones to
be taken to Triangle.

Three hours later we docked our little aluminum dinghy
beside the towering gray superstructure of the *Howay* in
Klemtu. The captain leaned over the railings. "Mr. Hancock, I

presume?" David greeted him with "Yes. Good evening, Captain Furlong. I'd like you to meet my assistant—my wife, Lyn."

His lower jaw dropped about three inches. "Is this your assistant?" he slowly asked.

It was obvious that David hadn't made it clear that his assistant was a woman.

7

Mother to a Murre

IT IS ALWAYS A PLEASURE TO BE ABOARD A FEDERAL FISHERIES boat. The crew are efficient and courteous, the officers welcoming and helpful. Captain Furlong, with clipped English speech, a friendly smile, and rosy cheeks, overcame his initial surprise regarding my sex and welcomed me aboard. While David concerned himself with stowing away the gear the captain showed me a fully equipped cabin with two bunks, writing desk and bathroom, then invited me to the officers' mess for coffee. Looking forward to being clean for the first time in weeks, I excused myself early and headed for the cabin. David was flabbergasted. "Where are you going?"

"To my cabin, of course. Goodnight," I replied.

In still horrified tones, David called, "Lyn, come back—that's the captain's cabin."

"But our things have been put in there and . . ." I continued to protest.

At this point Captain Furlong assured David that we were to sleep in his cabin and that was the end of that. Sometimes it is embarrassing to be a woman. The captain bunked in a small cubicle aft of the wheelhouse.

At 6 A.M., after a night of luxury, I looked through the port to see an imposing seven-hundred-foot hump of solid rock, the main distinguishing feature of famed Triangle Island. This island is the outermost of an irregular chain of islands, rocks, and reefs known as the Scott group that stretch their sawlike teeth thirty miles into the open Pacific from the northwest tip of Vancouver Island.

Even in summer the weather in this part of the Pacific can be forbidding. High winds up to 120 miles an hour, pounding surf and treacherous tides sometimes batter the treeless, fog-enshrouded rock and have taken toll of dozens of craft that now litter the ocean floor.

The home now of millions of seabirds nesting in the greatest concentration anywhere in the northeastern Pacific, Triangle once housed man.

In 1910 a lighthouse and wireless station were established at the summit, the highest light in the province. The remains of the wooden planking that supported the metal rails of a cable-drawn railway were now only a scar in the overgrown tangle of salmonberry. One thousand almost vertical steps led downward from the light to the beach.

Less than ten years after the light was built it was abandoned: it was so high it was almost continually obscured by fog and thus useless to shipping. It was so windy that the tower cracked, the roof of the living quarters, store, engine rooms, and store sheds blew off again and again. It was isolated and dreary and supplies were so irregular that lighthouse keepers would not stay. It wasn't safe for children to walk outside in a gale. The buildings leaked and it was often impossible to keep a fire going. In 1919 man left Triangle Island to the birds, seals, and incessant winds.

This was to be our home for the next month. The weather permitted our arrival. Would it allow our departure?

It was low tide. Our equipment had to be carried over three hundred feet of jagged rocks covered with slimy seaweed, then across storm-tossed driftwood of the boulder-strewn upper beach that butted sharply against the precipitous walls of the island itself.

We made our farewell to the crew of the *Howay*, who all expressed their delight that it was we and not they who were staying on the island.

Where to camp? The sloping grassy mountain rising sharply from the driftwood allowed no axing out of a campsite as on Solander. There was not enough flat space for a tent. We resorted to heaving the driftwood logs and washed-up planks into some kind of a floor for our sleeping bags. David rested a ridge pole between two huge boulders and battered logs and we threw canvas over the top and tied it down to the other logs. Although it seemed like taking coals to Newcastle we stowed our two thriving ancient murrelets inside the canopy with the equipment.

During our first night on Triangle the day's misty drizzle turned into a real downpour of rain. I felt like a bride snowed under white finery when David flung fifteen feet of cold plastic over our sleeping bags. How one's standards do change!

Next morning I was also covered with giant wood bugs, known more romantically as isopods. A song sparrow fluttered above our heads but it escaped under the flap of our canvas shelter before I could get the cameras organized.

For a brief reconnoiter of the mile-long island we clambered up the ridge above our camp and down the other side to a bay whose boulders had been completely blackened by a recent fire. During David's spring visit this half-mile beach had been piled ten to twenty-five feet high with logs thrown up by frenzied winter storms that had smashed against the westward side of the island. The entire beach had been clogged with logs a hundred feet inland. Apparently just after David's visit some fishermen had come ashore on Triangle and with a favorable wind lit a fire at one end of the beach. Soon the whole beach was ablaze. It must have been a grotesque holocaust, as millions upon millions of cords of wood went up in flames.

The strange aftermath of this conflagration now lay before us. Glass balls which had floated across the Pacific from Japan were intermingled with the charred driftwood as little globules in blue, green and brown. Landing on soft wood rather than on the rocks had saved them from breaking, but in the intense heat they had melted to cover the beach with small clumps of molten glass. Remnants of shipwrecks—iron fittings from hatch covers, bits of railings, and steel spikes—had drifted to shore. What saddened me most were the blackened bones of hundreds of seabirds that had once nested under the driftwood logs.

Climbing past the fascinating rock formations into the next bay I reacquainted myself with slipping shale affording little support for fingers and toes. Unlike Solander, Triangle lacks soil covering so that the short, stubby grass stumps break away at the touch and ominously roll down to the sea. David of course had found it all child's play and careered off to the sea-lion rocks. And eventually I caught up with him.

The rocky shores of Triangle Island itself are used by the Steller sea lions as hauling-out grounds, while those of the large islets lying immediately off the main island are breeding grounds and rookeries. We encountered far fewer animals than we expected. We saw only a couple of hundred sea lions around these rocks, whereas on previous visits David had counted up to eight hundred. Records showed that several thousand sea lions used to breed here. The numbers were dwindling not only on Triangle but on other breeding rookeries along the coast.

Closer inspection told us why. On government authority, a feed company had recently harvested many of the adult sea lions for mink food but only the hides of the pups to make into leather. Hundreds of their rotting carcasses lined the rocks. The tide pools smelled exactly like what they were—morgues. In one, eight skinned pups lay in a pile in pools of blood. We were overcome with the stench of death.

The actions of a few fishermen had compounded the problem. Annually hundreds of sea lions have been slaughtered by the kind of fisherman who looks for some excuse other than his own lack of ability to explain why he has done poorly at fishing. From recent studies we had learned that seventy-five percent of the diet of Steller sea lions consists of squid and octopus and the other twenty-five percent consists of bottom fish of little or no commercial value. Seldom does a Steller sea lion eat salmon. The complaints of the fishermen against the sea lion as competitor thus seem unjustified.

Fortunately the general attitude to our marine resources is improving. A few years ago the Federal Fisheries Department themselves launched annual sea-lion hunts in boats, armed with machine guns and heavy rifles. Hundreds of bulls, females, and pups were senselessly mowed down as they milled together on the breeding rookeries. Today a more enlightened approach to resource management prevails. The department now regulates how many animals can be commercially harvested for hides and mink food and discourages any wasteful destruction of these animals.

David has been concerned that the recent popularity of captive killer whales in aquariums may threaten the wild stocks. The Federal Fisheries Department has taken another positive stand in their recent protection of this magnificent animal. It now controls the number of capture permits issued.

But it seems ironic that while one can still kill a sea lion without repercussions, much bureaucratic red tape must be cut if one wants to keep a sea lion alive. This attitude isn't peculiar to the Federal Fisheries Department; it is an anachronism common to all regulations pertinent to wildlife.

Almost any individual can routinely buy a hunting license for from two to five dollars and go out in a season to kill a large number of animals. Very seldom can an individual get a permit to collect a single live animal. Normally one must belong to a scientific institution or a city zoo. If an animal is killed its value is limited and of short duration. If an animal is maintained alive it offers increased education and recreation to many people. Whereas an animal shot or captured alive gives almost equal returns to the economy, an animal maintained alive gives an additional boost by the necessity to house, feed, and maintain it.

A peculiarity of the law in British Columbia that disturbs me personally is the difference between the cost of a permit to kill an animal and one to take it alive. For example, a license to tag and kill a Dall sheep costs nine dollars, while a license to take one alive costs over two hundred.

Further idiosyncrasies trouble David. The licensed hunter —or should that be sadist?—can with full legal protection go out and kill a duck or mountain goat, then simply leave the animal to drift away on the tide or rot on a mountain slope; his

license permits him to kill but does not necessitate that he retrieve it or use it.

Lying crouched and silent behind a blind of boulders we
gazed fascinated at the remains of the colony of Steller sea lions.
The bulls are massive animals weighing a ton or a ton and a
quarter. Each harem master was surrounded by his harem of
sleek females, who seldom weigh over half a ton. Each male
holds a harem of ten to twenty females. The rich-chocolate-
brown pups, about forty-five pounds at birth, are not weaned
till they are around eleven months old. It is ludicrous to see a
male pup of about six hundred pounds still sucking on its
mother who is only a little bigger.

Normally the bulls remain in the middle of the colony surrounded by their females. The pups stay on the fringe so they
don't get squashed. A ton or more of blubber moving through
the colony will not deviate for smaller obstacles in its way.
Several golden-skinned mothers were playing with their pups
at the water's edge. One pup had remained on shore and its
mother was trying to coax it into the water. When both were
united they nuzzled each other and then frolicked off to another rock. Another mother seemed to be seal-sitting four
swimming pups at once.

Whereas fur seals like Sam are seldom found within five
miles from the British Columbia coast, Steller sea lions rarely
are found more than five miles from shore. In contrast also to
the long migration of the female and pup fur seals, many Steller
sea-lion cows and pups remain near the rookeries throughout
the entire year. Other cows and the bulls move from the exposed rookeries into the inlets or more sheltered bodies of
water. The migrations of these sea lions have not been fully
studied at present.

Triangle Island is unique. It contains the only ground-nesting bald eagles in the province. It has one of the two major sea-lion rookeries. It has the largest concentration of nesting seabirds in the entire northeastern Pacific, and the only colony of breeding murres in the province. Off the Santa Barbara coast in California during a recent oil spill about fifty thousand seabirds died. If an oil spill occurred off Triangle Island there would be a loss of five million birds. Unfortunately, in this isolated area, few would witness the tragedy and complain.

The only way to focus sufficient attention on this wilderness area would be to include all the Scott Islands in the newly proposed wilderness park for the northern end of Vancouver Island. Only then would sufficient antipollution regulations be enacted.

Around the corner from the sea-lion rookery we flushed a pair of bald eagles which were scavenging on the remains of sea-lion pups. Another pair, obviously hunting, flew low over the crown of the island. David has recorded four long-established nests on Triangle. But in four years only three young have fledged. He thinks inclement weather may be a factor in such poor reproductive success. On one occasion he found two dead sub-adult eagles floating in a shallow concrete-sided well at the base of the cableway remains.

On his spring visit earlier in the year he had built a blind eighty feet from an occupied nest to study and film the development of the young nestling. The day he finished building the blind it started to pour with rain. From his vantage point behind canvas he watched both parent eagles settle over the one young, using their wings to shield him from the rain.

After the rainforest jungles of Klemtu and Barkley Sound David was enthusiastic at the prospect of such close opportunities for unobstructed observation and photography. He set up camp in the blind but not once in the next seven days did the adult eagles return during daylight hours. They fed the eaglet at night and again in the morning before it was light. This arrangement was adequate for observations but not for photography.

David discovered that an eaglet in the nest does little throughout the day but grow. It preens its new feathers, plays with twigs, and backs up to the edge to fire its droppings over the side of the nest. I always think this is a rather ironic example of sanitation when the eaglet is sitting amid such piles of rotting food remains. In this particular nest David counted the remnants of several species of seabirds (rhinoceros auklets, puffins, murres) and eleven species of fish. This varied food supply suggests that eagles are primarily scavengers rather than hunters. Had they been actively hunting live quarry, they probably would have developed a pattern of capturing one or two species of prey. But a scavenger picks up whatever it finds available, and Triangle Island provided an abundant source of constantly dying birds.

We were intrigued by the wealth of life in the intertidal zone. Each tide pool at our front door was an aquarium of colorful creatures—vivid red, green, and purple sea urchins; orange starfish; yellow nudibranchs or shell-less relatives of the clams. The rocks were crammed with acres of giant orange California mussels and many different types of seaweed. Even when the sea was calm the waters still heaved slowly up and down, bringing new food, the plankton, to the creatures that clung to the rock surfaces. The rich upwelling of currents off the west coast supplies food for all this life.

For days our activities were dampened by a howling wind, drizzling rain that never ceased, and mists and fog that swirled around the cliff tops to completely obliterate the seabird colonies. We repaired our tent, fished off the rocks to feed our ancient murrelets, and collected specimens of peromyscus (white-footed mice) for the university.

I think for every bird on Triangle there must have been ten mice. If I left the lid off the water bucket some of the mice which had raided our larder during the night would be drowned by morning.

The mouse population of the island is unique. There are white-footed mice all over North America and on almost all the major islands around the continent. It is probable that more scientists have studied the white-footed mouse than any other animal. The Triangle Island mice are in great demand by scientists who wish to examine the characteristics that distinguish these large mice from those in other parts of the world. A further interesting factor is their presence on such an isolated island thirty miles from Vancouver Island.

One theory suggests that Triangle may not have been glaciated and the mice survived here through the Ice Age, a second that if the area was glaciated the mice ran over the ice and arrived on Triangle as the ice cap on the coast was melting. A third theory with perhaps more applicability suggests that the mice came down some of the coastal fjords and landslides, and then floated over to the island on mats of debris and tree roots that had sloughed down the sides of the slide. Still another theory is that they arrived by Indian canoe.

One of the early lighthouse keepers had released domestic rabbits on the island. Although rabbits are such prolific breeders and there was abundant grass available, we saw none on our current visit. Many people have commented on the number of rabbits seen at Triangle, but on only one previous visit has David encountered any. The Fisheries Department crew who were rowing him ashore on the occasion of his spring visit to study eagles insisted that the island was overrun with rabbits. David was plainly skeptical. No sooner had they landed the skiff than two rabbits went bounding up the beach and into the brush. Never again did he see a rabbit. David couldn't help but wonder if those two animals were just a pair of males!

Two comical oystercatchers lived among the boulders beside us. Sometimes their curiosity would lead them right up to the door of our shelter. We enjoyed playing peekaboo with them. Carefully following the rock contours, they sneaked closer and closer, poking their heads above the rocks to check whether we were looking before stretching their necks higher to improve their position for observation. If we ducked down out of sight they hopped along the rocks toward us on long pink legs and feet.

The oystercatcher's jet-black plumage, long bright-red chisel-shaped bill and golden red-rimmed eyes makes it almost as captivating in appearance as the tufted puffin.

For hours we searched among the bare rocks for signs of young. In contrast to the conspicuous adults, the mottling flecks of the eggs and young blend in perfectly with the patterns of the surrounding rocks. This cryptic coloring combined with the camouflage of absolute stillness is so effective that when I picked up my first chick oystercatcher I thought it was dead. It lay inert in my hand till David took it from me and put it on the ground. Immediately it made a headlong dash to the sea. While the parents hovered anxiously overhead it landed in the water a few yards from shore. Young oystercatchers do swim occasionally. Adults don't. We were surprised to see it dive just like a puffin or cormorant. When it finally broke the surface it was still completely dry. The adults flew off screaming and calling attention to themselves, even feigning broken wings, to lead us away from their offspring. We remained out of sight until the chick returned from the water and hid again in the rock crevices.

Another day we observed two tiny chicks hatch out of mottled eggs. When only a few hours old, they wobbled around like animated pin cushions and followed their parents down to the intertidal zone to learn how to detach limpets, chitons, mussels, sea urchins, and sea worms with quick flips of their long bills.

One morning I awoke to shake off the amphipods and isopods from my sleeping bag as a great glow suddenly illuminated the shelter. I rushed outside to see the wind part some of the clouds and reveal patches of blue sky and sun. Quickly David decided to establish our second camp on Murre Rock—our name for a secondary island about six hundred yards long, with very steep sides rising to a height of four hundred feet above the sea, but joined to the main island at low tide.

Loaded to my shoelaces with food, camera gear, and canvas for the blind, I groaned as the precipitous sides of Murre Rock loomed ahead. If Solander was a molehill, Murre Rock was a mountain. If I finally managed to get to the top, I decided, it would be better to camp there rather than descend to base camp each day, and then climb the cliff again the next morning.

A seabird colony is a beautifully complex structure. Generally isolated from destructive mammalian predators (except man), birds of many species and diversified requirements live together in an intricate web of conflicts and mutually beneficial interaction. Each species requires a slightly different type of habitat in which to nest. The oystercatcher chooses the open rocks just out of wave reach; the pigeon guillemot places its eggs under big boulders just above the high-tide mark; the glaucous-winged gull chooses the open grassy areas between rocks; the tufted puffin selects the earthy and grassy areas further up the cliffs, where there is sufficient soil to burrow; the murre balances its egg on a narrow ledge of the steep cliff face; while the cormorant builds its bulky seaweed-and-stick nest on wider ledges of less precipitous cliff faces.

The selection of varying nesting habitat is paralleled by a corresponding selection of feeding habitat.

The alcids around such a rookery as Triangle reveal some interesting feeding and behavior patterns. The pigeon guillemot feeds closest to the shore in the intertidal zone. It carries to its young only one or two small fish in each of its frequent trips. The murres, which feed a little farther out from the shore, make fewer trips back to the young. They bring only one fish at a time but it is larger in size. In both these species the abundance of food close at hand permits the rapid growth and early fledging of the young. Neither requires the added protection of a deep burrow for its young.

The tufted puffin feeds still farther out, so they make even fewer trips back to the young. To compensate, the puffin's beak enables it to carry many small fish at one time. Another alcid, the rhinoceros auklet, feeds in the same region as the puffin as well as even farther out. To avoid competition with the puffin it catches its food at night; it makes fewer trips than the puffin but carries larger fish. In both these latter species, the conditions of offshore feeding necessitate a slower growth and later fledging of the young, and both require the protection of a burrow for their young.

The diminutive Cassin's auklets spread themselves through the entire area; they avoid competition with other species by feeding on small shrimps, which they partially digest to concentrate the nutrients, and by flying to and from their burrows at night.

Our friend Dr. Martin Cody from Los Angeles has recently

been studying many of the west coast alcids, particularly off the Washington shoreline. He writes: "I believe that competition for feeding zones around a seabird island exists and that the resultant displacement patterns of feeding zones determine most of the remaining ecological-reproductive characteristics of the species."

This, David tells me, is a scientific confirmation by statistics of our own observations. Martin has made similar observations for the alcids of the Iceland coast.

A group of adult pigeon guillemots with bright-green blennies still wriggling in their bills were sunning themselves on the boulders to await our departure before flying into low rock crevices underneath to feed their young. I like the sleek black beauty and the brilliant red feet of the adult pigeon guillemot but not the pugnacity of their black, roly-poly, fluffball chicks with their ever-open needle-sharp beaks.

Above the boulders among the grassy humps that everywhere resembled long-haired shaggy mops, puffins had honeycombed the hillside with their homes. We found a tufted puffin that had launched himself headlong out of his burrow and lodged his leg in a branch of salmonberry. He was many days dead.

Overhead a glaucous-winged gull divebombed a tufted puffin as it returned to its burrow with a beakful of tiny herring, all neatly dangling by their heads in the powerful bill. The gaudy puffin did a quick sideslip, using its webbed orange feet as rudders to evade the raiding gull, and flew hastily into the safety of its nesting burrow. The puffin would then waddle four or five feet along an excreta-moistened tunnel to a cavity where a single downy chick awaited its dinner. The puffin must dig a burrow too long and narrow to be reached by the predatory gull.

Slowly we climbed upward. I pulled on roots and burrow roofs of grass clumps that I hoped would remain firm, to gain about two feet of height with each heave. We took frequent rests and enjoyed the only advantage of mountain climbing I could appreciate—the view. Sun and clouds alternated but the photogenic setting of the jagged hogback which ran in a semicircle around the picturesque bay improved with each step upward. I started with a shirt, two sweaters, and a ski jacket. I finished the climb with my body bare from the waist up, even if Scotch thistles took their toll from unprotected skin. The cool breeze was most invigorating.

"I like it, Lyn, but don't you think you'll get a cold?" David commented with a grin.

At the summit we lunched on raisins and delighted in the panorama of the Scott Island chain jutting upward from an unusually calm and silvery sea. Triangle, Sartine, Beresford, Lanz, and Cox, and even the smudge of Vancouver Island more than thirty miles to the east.

As we stood on the cliffs and looked inland, the grassy knolls of the island were like a well-kept golf course. But Triangle is fickle: one day she can be dressed in the sheer beauty of a blue glassy sea, a clear sky, a glorious sunset; next morning may bring gale-force winds, pounding seas, and rain squalls. Her golf-course meadows to the eye become up close a dense jungle of waist-high salmonberry that must be fought through without disturbing the petrels and auklets that make their homes in the tangled roots.

At the cliff edge on the other side of the island we located three murre colonies within camera reach on ledges about four feet from the top of the cliff, and another larger colony on a vertical rock face thirty feet down.

Murres are relatives of the extinct great auk and occupy in

the northern hemisphere the same ecological niche that is occupied in the southern hemisphere by the penguins, which they
resemble in posture and coloration. A nesting murre requires
scarcely one square foot of territory per bird. They are extremely sociable birds, breeding shoulder to shoulder in dense
crowds on the narrow ledges of exposed sheer cliffs. If a murre
nested out in the open by itself it would be very quickly over-
powered. Nesting in such density allows all the murres to put
their beaks together and growl to ward off a predator.

Carrying an oolachan lengthwise in his beak, tail sticking
out as a rudder, a new arrival from the sea braked hard with his
web feet splayed out to grasp the ledge, then flapped his wings
madly to maintain a hold as he jockeyed to his own territory. I
was baffled as to how he knew his mate or his own spot when
all the birds looked the same. Murres look like elegant men in
immaculate black tie and tails, dark chocolate brown on head
and back and pure white breast beneath.

His arrival caused much squabbling and increased noise. All
the neighbors growled and moaned and jabbed with their long
rapier bills in the direction of the newcomer, and the disruption
rippled along the entire ledge.

Murres beat their short, stubby wings very rapidly to give
an impression of speed and power, though actually they make
little progress against a strong wind and are easily outdistanced
by the gulls, which beat their wings at a more leisurely rate. A
murre literally dives into the air from its nesting ledge, hurtling
downward to the sea. In a few seconds it gains sufficient mo-
mentum for level flight. Returning to its nesting ledge at a
higher altitude, it avoids collision with any murre seabound.
When emerging from the water, a murre becomes airborne
only after a long run, using the crest of the wave as a trampoline
and striking the water with its wings to accelerate its speed.

The short, narrow wings of the murre are adapted for under-water locomotion. They act like powerful paddles to aid the muscular legs to propel the bird through the water.

We were to collect young murres during our stay on the island, so we studied the colony carefully for eggs and chicks. Mr. Conway of New York wanted large, downy young, but all we could see were eggs. The adults who were incubating them busily attempted to turn and delicately resettle their beautiful brown-and-green-blotched eggs under their white bellies for warmth.

Not only does the murre produce one of the most attractive eggs in the bird world but it also takes first prize for an egg which is unique in shape and function: large at one end and tapering to a point at the other. When the egg rolls it spins in a tight circle, reducing the danger of falling off the narrow ledge. Nevertheless the loss of eggs in a colony is substantial. Probably no more than half the eggs hatch.

Chick murres spend eighteen to twenty-five days on the ledge before fluttering down to the sea on their own. At this time the adults congregate in a milling mass at the base of the cliffs, calling and growling excitedly. The chick swims close to the adult that most persistently answers its calls and is led away to the sea. In our many miles along the coast we have never seen a chick accompanied by two adults.

The basic problem in a seabird apartment house is the plumbing. He who nests at the bottom of the colony bears the brunt of the housecleaning actions of the ones above. A couple of murres seemed to have been splashed upon for days. Gray and white excreta streaked and spoiled their beautiful sleek plumage.

Above the murre colony perched the more dignified pelagic cormorants on the edges of their crowded nests. The adult cormorants, gular or throat pouches continually throbbing, seemed oblivious of ugly, naked little nestlings similarly throbbing up and down at their feet like snakes under the influence of a charmer's music. At times the mother was stimulated to lower her beak so an insistent offspring could thrust its head and the full length of its immense neck down the parent's gullet in search of partially regurgitated fish. With its bill filled and its eyes and face covered with the half-digested slime Junior flopped limply back to the nest to rest and pant in exhaustion.

Gulls seemed to have nothing to do but sit and scold or glide and scream, while the speckled, fluffy young lay under grass clumps between the puffin burrows. The gull is the main predator of the seabird colony. The bald eagle is a less common predator.

We organized our cameras and concealed ourselves in a blind just in time to film a bald eagle flying above the murre colony. As the eagle passed over, hundreds of birds launched themselves seaward, leaving many eggs and chicks unguarded. The eagle casually sailed on, uninterested in the thousands of birds that panicked below him.

At the instant the eggs were left unattended by the cormorants and murres, the gulls dived and swooped down. At one temporarily deserted cormorant nest a gull landed to gobble down three eggs in less than three seconds.

With the murre ledge newly abandoned we could now more easily check the number of eggs and chicks. Only one chick was revealed. Just as we were commenting on its diminutive size, a second gull landed on the ledge and grabbed it. It was swallowed whole.

On an adjacent rock promontory another gull sat on a cormorant's nest that contained two one-quarter-grown cor-

morant chicks. The predator grabbed the closer one and repeatedly turned it over in its great beak to organize its position
more thoroughly. With a quick gargle it attempted to swallow
the chick rear first, but the large stomach prevented the com-
pletion of the action. The gull dropped the chick, turned it a
few more times and tried to swallow it headfirst. This action
went well until the large abdomen caught and blocked the gape
of the gull. The chick's feet dangled and twitched from the
corner of the gull's mouth. At this point the gull simply
squatted and rested on the nest. Perhaps he would have stayed
there for some time had it not been for the sudden return of the
parent cormorant. Before the gull could rise from the nest the
cormorant had jabbed it twice with its ferocious hooked beak.
The gull disappeared with the cormorant chick still protruding
from its beak. Disregarding its brother's fate, the second chick
adopted its food-begging posture, the adult resumed the brood-
ing position over its lone youngster, and the colony resumed its
normal activity.

This example of gull predation vividly brought home to us
the havoc people can also wreak in a seabird colony. We remembered to tread carefully to avoid caving in the soft earth
roofs of the nesting tunnels, to prevent crushing eggs or young.
We were more aware that the greatest hazard to nesting birds
is indirect. When a young gull chick is disturbed it runs away
from its nesting area. Once displaced, a young gull's chances
of survival are small. Adults usually kill all foreign chicks in
their territories.

As darkness was imminent we stopped birdwatching to set
up the small tent. In order to disturb the fewest-possible number of birds yet use the tent as a blind to overlook the murre
colony, we had to camp on the very edge of the cliff. I dreaded
the effect of increasing wind on my fear of heights. I positioned

the gear along the cliff side of the tent to provide some anchorage for the canvas in case a gale sprang up during the night. Once inside the sleeping bag I tried to forget how close I was to a four-hundred-foot vertical drop.

The bombardment of the auklets that flew against the sides of the tent, the cries and scoldings of the gulls, the homecoming cries of the petrels, the pattering of feet in the ancient murrelets' box, and the hard stony ground made our first night on top entirely sleepless.

We spent the next morning in the tent observing and filming the puffin burrows directly in front of us and the murre colony just below.

As on Solander, the puffins sat socializing on the grassy humps in front of their burrows, their golden tassels flying backward in the breeze, yawning, closing their eyes, stretching, ruffling their feathers and settling down again. When wind currents invited, they zoomed down over the sea and circled busily. The gulls, in contrast, with wings outspread and motionless, floated effortlessly with each up and down draft.

Light conditions for photography were frustrating. In the swirling fog the rocks and cliffs of the background alternately appeared and disappeared. Within seconds light exposures altered, then altered again before a shot was taken. One minute I was bare to the waist in the heat and the next I pulled on a sweater against the wind.

During the afternoon we emerged to investigate this side of the island more thoroughly. David hoped to select a small colony of isolated murres for the eventual collection of the chicks, rather than cause disturbance to the main colonies.

While David pushed ahead in his own investigations I ambled along the top of the cliff on the outskirts of the puffin burrows and tried to accustom myself to a bird's-eye view of the foam-patterned, surf-curled sea. I was surprised to see a couple of murres standing among the puffins, obviously out of place. As I approached, the adult birds plunged seaward, to reveal one beautiful olive-green egg delicately mottled with chocolate and a perky young murre chick, the fluffy image of his immaculately coated mother. I named him Perky as he waddled jauntily around a puffin hillock and allowed me to pick him up. I planned to pretend to David that I had lowered myself over the cliffs to secure an egg and a chick.

I don't think David shared my enthusiasm when I sheepishly approached him with my sweater bulging. Suspiciously he looked at my gyrating third breast and was about to open his mouth when Perky struggled up from my bra and popped his head out of the V in my sweater.

"What are you planning to do with him? We don't need to collect our murres until the last day of the trip," he reminded me, trying not to laugh.

"But all we have seen are *eggs*. Surely when they hatch they won't be the size we need in time for when we leave. Why don't we try to raise a couple of them ourselves as an experiment—in case we have to do that with the lot when we leave? We can put Perky with the ancient murrelets. You always wanted to study the murre's development from the egg stage onward. I brought you something else too. . . ."

At this point urgent peeping sounds were being emitted from my bosom, where the egg was being incubated, so I thought I'd better confess immediately.

David smiled. "You forgot one thing, my love. That codfish we brought up with us for the ancient murrelets will not be sufficient for all the chicks we're going to have in the tent tomorrow. This egg will hatch tonight. See, the large end is already cracked and pipped. I'll stay up here and film the murre colony while you take a short trip to the bottom to catch some fish and haul up some sea water for your little brood."

I groaned, "Oh no! I vowed when I got to the top I wouldn't be returning back up again. Give them back to me. I'll take them back."

"You'll never put them back in exactly the same place. A murre doesn't really recognize her own egg but she does know the nest site. We might as well look upon it as an experiment and we'll both descend the cliff tomorrow," David consented magnanimously.

We returned to the tent, where feeding duties provided the entertainment for the evening. I kept the insistently peeping egg under my sweater. Perky didn't seem too hungry, at least not for cod. He was far more interested in stretching up proudly to his fullest height, shaking his tiny wing tips and waddling around on explorations of the tent. He stood head and shoulders above the murrelets.

As I popped into my sleeping bag I asked casually, "Dave, where do you want the egg now?"

David protested, "But, love, this is a woman's job. Men don't know anything about bringing babies into the world."

"If I have it in my sleeping bag I'll crush it during the night," I said firmly, passing along the egg to the biologist.

David wrapped Perky in a sweater near his pillow so that he could hear any distress calls if it got cold. He then placed the little Peeper in a towel next to his chest.

Night was once again nearly sleepless. We became so absorbed with the fight for life inside the limestone womb that the hours passed quickly. I held the torch while we watched the egg tooth, that tiny hard calcareous tip at the end of his beak, work its way slowly around the pipped section. Just before midnight the egg tooth had completely rotated around the top of the egg, freeing the cap to permit the exit. This gradual wearing away of the cap had taken nearly two days of constant trouble. Air was now entering his lungs and it was as if he were trying to expand them to their total volume to force the cap away. His weak struggles and peeping made me want to help him but David said that interfering could only be detrimental. If we peeped, Peeper replied, and immediately began a new burst of wriggling. When David thought his rest periods were too long he would peep again to stimulate activity.

To David's disgust I soon fell asleep. But at 4 A.M. he nudged me awake just as an extra-large inhalation and kick pushed the cap free and a wet little infant murre emerged, with black scraggly body, tiny crooked arms bent at an odd angle and a black-and-white hairy head like a velour cap. His abdomen was now swollen with the last of the absorbed yolk that had been his food source. As I watched in fascination David detached the sac containing the waste products of his metabolism and returned him to the warmth of the sleeping bag to dry and rest.

A few hours later in the light of day we investigated him more closely. Now dried out, he was black fluff above and white fluff beneath. His baby calls were just as persistent as when in the egg and we debated how to answer them.

Dr. Konrad Lorenz, a famous German student of animal behavior, discovered that some newborn animals accept the first living thing they see as their mother whether it is their natural mother or some other animal. This Lorenz principle is known as imprinting. We now had the responsibility of being the mother and father of a murre.

If we kept Peeper alive, we could collect eggs and hatch them. Not only would we fulfill our promise to the New York Zoological Society, but our own research on the development and behavior of murres would be more complete.

The joys of motherhood were considerably lessened when a baby murre pooped with regularity between my shirt and bare warm skin!

I sat for hours trying to persuade newly hatched Peeper to gulp down a piece of fish but David said I was acting the overprotective parent and that normally the young chick wouldn't eat for the first twenty-four, thirty-six, or even forty-eight hours. During that early period it would live on the yolk sac which it reabsorbed when it hatched. However, I dangled that fish in front of him for so long that he must have got tired of seeing it, for suddenly he swallowed it. It wasn't easy, and he had frequent rests before the long strip eventually disappeared. An adult murre would have brought her young a fish perhaps four inches long, and he would have had to digest the head before he could digest the belly.

Without gear, the descent from Murre Rock was much easier than the formidable ascent. My method is hands and knees upward and bottom bumping downward, resulting of course in a soaked seat.

Despite the heavy surf and swirling seas, with much perseverance David managed to catch four codfish. With the addition of limpets and vitamin drops I hoped the cod would provide sufficient food value to equate with whole sandlance, needlefish, oolachan, and blennies that the wild bird would have been fed. My job was to bring in the fish through the crashing waves with the net after David caught it with the rod and line. To carry it, I then threaded my shoelaces through the gills and those ghastly thick cod lips. Only a mother could love a codfish!

My first efforts at filleting and gutting were achieved with the help of an axe and gloves.

At the top of Murre Rock, lack of water was a problem.

Our chicks couldn't swallow the coarse strips of cod without some sort of lubrication so we tried soaking the fish in a can of ginger ale. As you might guess after each meal they burped!

I had insisted on cutting down utensils to a single knife, because utensils were part of my pack. A knife already blunted by engineering duties during tent raising and then covered with codfish innards and limpet juices after I had filleted and cut up fish strips for the chicks just couldn't be used to cut our own rations. So for breakfast we had boiled eggs broken up by teeth and spat out on bread.

It always amazes me with a slight degree of envy that other outdoor wives appear immaculately groomed when they are so far removed from the comforts of home. Keeping clean under a constant shower of seabird droppings on an island that lacks fresh water is a special challenge. After rooting my way up a mountain on my hands and knees I can usually eradicate a little dirt by rubbing my hands through the soil and wiping them afterward on the stiff grass of a puffin hummock. I have tried makeup as a cover, but the result was always too hilarious and I always decided to wait for hot water and scouring powder on my return to civilization.

Most of the chicks in our menagerie were eager little feeders. No sooner did I show the food at the top of the box than they jumped to gobble it down. Yet still there were individual differences. The ancient murrelets were the quickest to whisk away the proffered food. Perky, the murre chick, seemed to eat primarily in anger, defiance, or to prove superiority.

The major problem was to keep the boxes clean and free of droppings as far as possible. When their feathers were fully developed at home they would gradually regain their waterproofing. On Murre Rock I replaced fresh grass from the surrounding puffin burrows as many times as needed in the day.

For the ornithologist Triangle Island, harboring the largest breeding population of tufted puffins, Cassin's auklets, and common murres on the British Columbia coast, is a bird paradise. Although during the day gulls squawk, cormorants croak, guillemots whine, puffins growl, and falcons kack at chittering bald eagles to the accompaniment of a constant wind moaning like the chorus of a howling wolf pack, most of the activity and bedlam is nocturnal.

One night David made his greatest discovery. Although no other previous scientific survey of Triangle Island had recorded the presence of rhinoceros auklets, David felt sure that the salmonberry-covered southern slopes of the island should house this bizarre bird, as well as the thousands of Cassin's auklets and petrels we found by identifying calls and sampling burrows.

In preparation for the night flight we strung up the mist net across typical rhinoceros auklet habitat, then sat in the dark among the salmonberry thickets. We didn't have long to wait. Birds that had spent the day at sea hurtled in to incubate and feed their young in the burrows—and every bird that hit our net was a rhinoceros auklet. We were constantly on the run, scrambling up and down through the tangle of salmonberry in the darkness to extricate and release the auklets from the net. The whir of countless wings made a solid background of noise. David collected a small live sample selection from divergent burrows to study at home and we stumbled back to our tent on the logs. Despite aching bones and stiff joints I tried to share David's enthusiasm in locating such an extensive continuous colony of unrecorded rhinos. Each auklet had to be kept separate from the others, and to prevent self-inflicted damage David placed each in a sack till they settled down.

In the light of the moon I studied this further member of

the alcid family. Rhinoceros auklet is aptly named for the large horny projection on its top mandible. The long, narrow beak is paler orange than a puffin's but almost as powerful. The conspicuous facial plumes or whiskers are shed in the winter, as are the tufts and the beak sheath of the puffin.

The waves of greatest rhino activity occurred just after dusk and again just before dawn. Most of the incoming birds carried a single fish in their bills. Incredibly, one or two fish from each parent are sufficient to sustain a young chick, though I had observed other alcids bringing a dozen or more (admittedly smaller ones) to the burrow in the daylight. And even more amazing is the distance the birds travel from their feeding grounds to the nesting burrows.

Next day we examined the colony more thoroughly. Most of the burrows were very deep and many of the tunnels were forked with numerous blind side alleys, indicating a long-established colony.

Well-worn burrows and moist fecal matter indicated occupancy. One burrow tunneled in three feet and then veered left. After eight feet more and forty minutes of cutting salmonberry roots it turned right and at the end of this was the enlarged terminal nesting chamber—empty. Already the young rhino had been abandoned by its parents and had ventured into the sea to earn its own living. This was the exception. At this time in early August seventy-five percent of the young were between one-half and two-thirds grown.

The remaining days were hectic as our observations and our filming continued. Every day was so packed with activity and interest that time almost faded into insignificance. Our expedition to Triangle was almost jeopardized because we couldn't remember exactly what day it was.

Although I thought I'd been consistently reporting a daily

diary, events had usually outdistanced my writing about them. David thought it was Friday: I thought it was Thursday. It was important to know exactly the day of departure as it was essential to collect the murres on the very day we were going to leave the island. Since several hours were required to get the birds from the top of Murre Rock to the base camp, timing was very important. The *Howay* was to collect us later on a Friday. Was this Friday? My diary indicated it should be Thursday. But had I missed a day?

Eventually after much argument back and forth we had an unexpected chance to test each other's hypothesis. For the first time in a month a trawler entered the bay. We quickly worked down the cliff and across the slippery rocks to get within yelling distance.

"What day is it?"

Silence. He dived inside his galley. Minutes later he emerged.

Above the wind came "Friday."

So we climbed the mountain again to start our collecting of seabirds. Still at the back of my mind I had that nagging feeling it was Thursday. But remarkably we had a second chance to dispel my lingering doubts when a second trawler entered the bay and a boy rowed ashore.

From halfway up the cliff David decided we should again descend to base camp. And after the usual greetings and expressions of surprise to find each other in such a remote area, I casually whispered to the boy in the boat, "You don't know what day it is, do you?"

"No, but I'll find out when I go back on board. I'll tie an orange jacket on the bow if it is Friday."

Anxiously I gazed at the trawler as she chugged out of the bay. Gleefully David shook me. "See, orange jacket on the bow. Silly! It *is* Friday. Two other people and I can't be wrong."

The rush was on. The weather pattern had improved and there was little chance that our departure would be delayed by sudden storms. We dismantled both camps, cached the unused food and assembled our gear. Now all that remained was to collect the murres.

By this date several chicks had hatched along the isolated narrow ledge that we had selected for our collection. I was no help at all in catching murres. Nothing would persuade me to walk along the edge of that vertical cliff, let alone lower myself over that abyss of blue crawling water or even to swing the dip net over and down the cliff to surprise the adult murres roosting on the ledge.

In far less time than when he had caught puffins on Solander, David had netted our permitted number of sleek, shiny adults. To make certain that no egg or chick would be parentless by our choice of adults we removed all the eggs and chicks on that particular ledge.

Taking care not to ruffle the feathers and interfere with their waterproofing we placed the adults in the grass-padded carrying crates. I carried the chicks and eggs in an emptied camera case. The eggs were packed in grass at the bottom, then the chicks were placed on top of the eggs to add heat for incubation. David then struggled downhill with the boxes strapped to his rucksack.

As we sat on the site of our dismantled camp to await our boat we only regretted the fact that we hadn't been able to make use of the day's unbelievably perfect weather for color photography.

The sea was calm and the sunset beautiful. Where was the boat?

Our eyes met and the terrible suspicion grew into a certainty: two trawlers had been wrong. My diary was right. It wasn't Friday, our last day: it was Thursday. We were ready to leave, but our boat wasn't due till the following day.

We spent Thursday night more uncomfortably than ever in our sleeping bags on the driftwood logs.

The next day also was ideal for photography but we were too engrossed from dawn till dusk in catching enough fish to feed our crated charges.

Finally the sleek gray lines of the Fisheries Patrol vessel anchored in the bay and the small skiff put ashore. When First Officer Arthur Tulk greeted us we realized that it wasn't the *Howay* that had come to collect us but her sister ship the *Laurier*. Normally this would have caused no concern. However, David had given the *Howay* a hundred pounds of frozen herring. Now the *Howay* with the herring on board had gone to chase some Russian trawlers.

After the skiff returned to the ship with our first load of gear and the news of our need for fish, the crew threw their lines over the side and jigged for cod.

We had intended to clean up and change into another set of clothes before we returned to civilization, but as usual time did not permit and Captain McLellan greeted two scruffy, tired, and dirty Hancocks as we climbed aboard with the last load.

One job remained to be done. We had a permit from the Federal Fisheries Department to collect a Steller sea-lion pup from the Triangle Island rookery. We had two reasons for wanting one. Sam had spurned all attempts to give him a harbor seal as a friend. Perhaps he would show more interest in the more closely related Steller sea lion. David's second reason was to continue his experiments on raising seals and sea lions. Only two or three Steller sea-lion pups had ever been reared. They were rarely kept in captivity and even more rarely weaned. One of the main problems is that sea lions contain very little lactase in their bodies. Lactase is the enzyme which is used to break down lactose, the main sugar component of mammal's milk. In order to obtain energy from mammal milk, lactase must be present.

Sea lions have a completely unique form of metabolism, which is based on breaking down the fats of their very fatty milk. Sea lions need to have very rich milk. Pups have to endure long periods of fasting when their mothers leave them for many days to go to sea and feed. The fur-seal pup may have to go two weeks without being fed. To overcome this long period of fasting they must be filled with very highly nutritious milk. The stomach is not large enough to store enough of the ordinary kind.

People tend to raise animals in captivity on cow's milk, or canned milk which is a derivative of cow's milk. But if cow's milk is fed to a sea lion it just passes through him and he derives no nourishment. In effect you can fill the animal with food and he will still die of starvation.

David wanted to try some ideas he had for feeding a sea lion directly on ground-up fish, bonemeal, and fish oil. He felt that if the animals eventually were able to break down fish they must already have in their bodies the necessary enzymes. He intended to feed the sea-lion pup on very finely ground fish to enable the stomach to handle it.

As the *Laurier* had other commitments only one hour could be allotted to the capture of a pup. All on board said it couldn't be done. The waves were high, the fast current surged violently over the reefs bounding the rookery. It was dangerous to take a small boat among the treacherous rocks. Dangerous animals were waiting on shore.

Only Art Tulk, the doughty Newfoundlander, offered to accompany David in the small skiff.

I wanted to film the experience, but now that I was with men I had to act like a woman, so I remained on deck and stared over the railings as David and Art cruised through the reef across to the sea-lion rocks a few hundred yards away. I wondered how they would handle the landing in the fast swells. Art would have to stay with the boat while David jumped ashore to scoop up a sea-lion pup before it scrambled into the sea.

Twenty minutes after they had left the *Laurier*, Captain McLellan lowered his binoculars to inform me incredulously that David was rolling up a sea lion in a tarpaulin and that the boat was on its way back to the ship.

In honor of the captain we named the forty-five-pound ball of belligerence Reginald. He was stowed away in the stern while we gorged thankfully on the cook's roast chicken, gourmet fare after our rations of the previous month. Reluctantly we turned down the long-anticipated hot baths as the birds had to be fed, the eggs incubated, and the chicks brooded.

After dinner it was all hands on the afterdeck for feeding. The Chief Engineer remarked that it was lucky having a woman on board. Otherwise he didn't see how the crew could have been so cheerfully organized as each came off duty. Walter the cook filleted cod, Arthur cut it into strips, Peter passed along hungry birds. John held the ones that were fed. Some scrounged new boxes, other cleaned dirty ones.

If I thought I had been a broody hen on Triangle by keeping eggs and chicks warm under my sweater I was far surpassed on the *Laurier* by Jim, a plump, jovial fellow with a wide grin who stuffed away in shirt, under sweaters, and into pockets every chick that had been fed. David suggested hiring him the following summer as an incubator. As hatching was imminent in several of the pipped eggs Brian, the oiler, offered to keep them near his pillow in the heated boiler room.

Feeding was an endless chore. The cod jigged by the crew didn't last long and again the crew came to the rescue. From Art a frozen ling cod, from Pete a sockeye salmon he'd intended as a gift for his mother, from Walter two quarts of herring. I was oblivious of the journey down the east side of Vancouver Island toward Vancouver: the deck was my whole world. After the last feeding the boxes were closed securely and marked clearly for their journey to New York.

We intended to raise the eggs and smaller chicks at Island View Beach. Seeing all the boats and fishermen enjoying themselves on a warm holiday afternoon in Georgia Strait as we neared Vancouver, I suddenly realized how hot it was. I had been too busy to remove all those Triangle Island sweaters.

As soon as the *Laurier* docked David said goodbye to those on board and rushed the birds to his parked car for the journey to the airport, while I continued to the *Laurier's* home base in Victoria.

Imagine my surprise when there on the dock was David! The explanation was typical: he got to the airport to find there was an air strike. No aircraft had left Vancouver for four days. Everybody on board had been so engrossed with the birds and we'd been too busy to listen to the radio; no one had thought to tell us about the strike. So David had transshipped the birds by the faster ferry to Victoria. Later when the air strike was over they would continue to New York.

It was an exceptionally informative expedition and it was only the beginning of the challenges that were to accompany the rearing of all our charges. During the next ten days at home every egg hatched. The excitement we felt and the satisfaction we received were in part a reflection of the earlier reaction of the oiler, who, as the *Laurier* was passing under Vancouver's Lions' Gate Bridge, burst up from the engine room, his arms flailing, calling out, "I'm a father, I'm a father . . . !"

To raise the young murre chicks as naturally as possible, David built a rocky clifflike ledge for them. Perky, the largest murre chick, was used to brood each hatching chick. The ledge was an advantage in that each chick waddled backward and voided over the edge, thus keeping itself clean.

As Lorenz had discovered, newly hatched goslings will persistently follow any large moving object, including the experimenter himself, and after a few days will fail to respond at all to birds of the same species. My murre chicks responded to me as their mother. Perky the eldest was the first to jump down off the ledge. Immediately she followed me as her "first large moving object." As each in turn jumped down it would follow me in long single file through the raspberry bushes and around the garden—twelve immaculate little men in black tie and tails. Unlike the adult, chick murres which have not yet entered the water can walk normally, with the joints of their legs moving freely. They stretch up tall on thin, spindly legs, whereas in the adult the leg has developed as a highly specialized swimming organ and the bird on land waddles clumsily and flatfootedly on its webbed feet.

Reginald arrived in Victoria wrapped in a tarpaulin. Emerging from the folds of the tarp with the typical Steller-sea-lion growl, which sounds like a human retching, he surveyed the backyard, which was to be his home for the last three weeks of the summer before our return to Vancouver. David had made friendly overtures to Reginald while his head was protruding from the tarp. It was a little difficult to do so now that he was careering around the lawn.

Although he was a roly-poly tub of lard we lost no time in trying to feed him, and David hoped to raise Reginald on an immediate fish diet, for the reasons I mentioned earlier.

My blender became the center of operations. We cut up all sorts of fish—salmon, cod, rockfish, sea perch and herring— then mixed them in the blender with a little water laced with multivitamins, minerals, salt, extra vitamin B, fish oils, and fish-bone meal.

Dressed in a heavy leather jacket and gauntlets, David caught Reginald with the salmon-landing net, then wedged his squirming bulk between his knees. David's mother was pressed into service to kneel over Reginald's wriggling rear to keep that in place while I knelt in front ready to pump in the fish. David stuck a piece of wood between Reginald's jaws while I thrust the long rubber tube down into his stomach. It was important to keep the blended mixture at the correct angle for pumping into his esophagus rather than his trachea. Once the tube was in the right position I could easily pump in the whole quart. Reginald's stomach bloated visibly underneath David's knees.

Day after day we chased Reginald around the lawn to intubate him with up to three quarts of the viscous mixture. With the formula and method perfected, David left again for Klemtu and Reginald was left to the folks and me.

As soon as he heard the blender and saw the pump Reginald slithered to faraway raspberry bushes. Why didn't he realize that these instruments of torture were the cause of the warm, satisfied feeling after each intubation? Our rides around the lawn did result in Reginald's gain in weight of two pounds a day and his elimination was normal so we knew his stomach was digesting the fish.

Reginald, being only a couple of weeks old, had mere bumps for teeth. Nevertheless at forty-five pounds he was ten pounds heavier than three-year-old Sam and could pack a bruising punch when he threw himself at someone. At maturity Reginald would weigh a ton, Sam only a diminutive six hundred pounds.

With David still studying eagles during the salmon-spawning season at Klemtu, I returned to Vancouver with Reginald, Sam, and a few of the seabirds that hadn't been shipped to New York. Our landlord and the neighbors now had to get used to two sea lions in our backyard.

On the first day of school I had a phone call from David in Klemtu. After the usual greetings I began to detect that David was forcing himself to be overly cheerful. In the back of my mind I felt that he hadn't phoned merely to tell me Klemtu was enjoying nice weather and that there were even more eagles around, and I finally asked what was wrong.

"Nothing's really wrong. I'm fine but I am afraid our little Piper Pacer isn't very well."

"David! What happened to the plane?" I asked frantically, thinking of flying into an eagle, or engine failure.

"Well, it burned to the ground," he answered so casually that I could tell how upset he was. "Nobody was hurt. In fact it was tied up at a float, but it turns out that the whole of Qua Qua camp burned down. Nobody was able to cut the aircraft loose. So it's a very sorry plane at the moment. It's still tied up at the float. The flames blazed off the cabin, caught the fabric and melted the wings. The engine of course burned up, and got so hot that it burned off the top of the floats. But sitting behind the float as it was, the waves hadn't drowned it. It's still floating, but beyond repair. I phoned the insurance man; he's on his way to investigate it now."

I felt sick. I could hardly bear the thought of losing our float plane. Memories rushed through my mind of our engagement flight to count eagles in the Gulf Islands, the first happy summer in Barkley Sound, our flights around Klemtu, where the work was only beginning. . . . I changed the subject.

"When are you coming home?"

"I hadn't really wanted to come back at this point. I've still got a number of things to do. I was going to spend another ten days up Laredo Inlet watching the eagles in the Bay of Plenty, which now has a large salmon run. I'll do that if I can get a boat. Then I'll have to charter an aircraft for the final surveys in this area. By then all the young will be out of the nest. I'll come down by mid-September."

Reginald now weighed eighty-three pounds. No longer did we pump in the food. Now that he could digest the ground fish we tried to get him to suck on whole herring. He would watch Sam gorge on twenty-five pounds, he would suck on the water in the dish or he would suck on his own flippers as a baby does, but never would he swallow a whole herring. Had it not been for the memory of our struggle with Sam and the eventual success, I would have long ago given up hope. Forcing down herring in quantity was infinitely more difficult than intubating.

Reginald was fatter and putting on double the weight that he normally would have gained in the wild. Upon his return, David decided to stop feeding him for a short period. His bellowing each night kept the neighbors awake and he still hadn't recognized fish as food by the end of the week, so we returned to force-feeding him whole herring. In a fortnight's time we tried another week of starvation. He lost ten pounds but he refused the herring as adamantly as ever.

Despite the daily force-feeding period he was no longer belligerent. In fact he was surprisingly tame. He flippered into the study and hauled his seventy-five pounds up onto David's lap, threw his front flippers around David's shoulders, his long whiskers around his face, and finally went to sleep on his knees.

One day he came into the kitchen while I was rolling out pastry and tried to heave his bulk from the floor up onto the pie plates. Not as agile as Sam, he fell with a thud to the floor.

Sam found in Reginald the ideal playmate. The two young sea lions became very much attached to each other, frolicked together, slept together, rode in the car together, but would not eat together.

We resorted to feeding Reginald with the tube. The problem was now quite serious. He had dropped weight now for more than a month. Yet he still showed no signs of wanting to eat on his own. He would gain two pounds of weight every day he was force-fed, drop two pounds every day he was not force-fed. We had to balance it back and forth.

Finally we went out one morning to find that he had lost his ebullient spirit. Despite an immediate dose of antibiotics his condition deteriorated over the next two days and on the third morning he was dead.

The post-mortem showed that at some point during intubation we had pushed the tube too far into his stomach. It had partially ruptured the intestine or the stomach wall and an infection had grown.

Despite the weeks of frustration of feeding and the failure in getting him to eat on his own, the loss was deeply felt.

Especially by Sam. He moped around the yard for the next few weeks like a lost soul.

8

A Home at Last

I WISH WE HAD A PICTURE OF THE HANCOCK MENAGERIE AS we left Vancouver the following summer to live permanently at Island View Beach. Hilary had arranged a farewell dinner party for us but we had to cut it short and leave for the island during the evening to avoid subjecting the seabirds to the heat of the following day. Our borrowed truck drooped to the ground under a bulky cargo of chairs, kitchen table, chesterfield, trunks, crated puffins and murres, a couple of boxes of herring and Sam the seal.

From dawn to dark during the following month we worked at a frenzied pace to raise the level of our land by trucking in sawdust and gravel. The vast pyramids of sawdust and mountains of gravel were flattened by rake and shovel in the hands of visitors who came to gaze but stayed to work. Relishing any activity in the open air, Hilary came to help on weekends. Both being sun worshipers, we stripped to the skimpiest of swim suits and then not always those after the ten-foot-high fence was completed. We worked busily at building pens as well as suntans until we heard one of the neighbors had a telescope trained on us.

We dug an extensive salt-water pool and walk-in aviary for Sam and the seabirds, then snow-fencing enclosures for the raptorial birds. Now with a place of his own, David planned to intensify his studies of endangered species like the bald eagle and the peregrine falcon.

Sea water had to be constantly circulated in all tanks and pools to provide marine mammals and birds with an oil-free surface. It was months before the system was perfected and then we were at the mercy of the tides.

First we blasted a trench fourteen hundred feet out to sea beyond the low-tide line and laid a plastic pipe. When water first trickled through the pipe we felt more triumphant than if we'd struck oil or gold. Just when we felt most elated the water stopped running in from the sea, the birds began to lose their waterproofing, and the pumps stopped running, not once but countless times.

Frustrating days were spent checking everything at both ends of the system. In most cases the trouble was not in the pumps and the valves but the pipe in between. Sucking in sea water over such a long distance caused the sides of the pipe to cave in. Many weeks and many problems later we hired a back-hoe to dig up the trench to the sea as far as the lowest tide allowed and replaced the plastic pipe with a metal one.

Despite a summer spent flattening sawdust and gravel, mixing cement, digging ditches, tearing down old sheds and building new ones, painting, wiring, and cleaning, I began to appreciate life at Island View Beach. The summer was hot and dry. Each morning a huge orange shimmering ball rose behind the islands from a mirrorlike sheet of sea a few steps from our front door and bathed the house in golden light. A great blue heron stood motionless above the newly exposed tide pools. A pair of loons glided with dignity past the boat ramp.

The stretch of water between Island View Beach and the San Juan Islands on our horizon is a living laboratory of bird life. Several pairs of grebes worked daily up and down the beach in their search for fish. Several hundred pairs of rhinoceros auklets that probably nested on a nearby island in Washington fed on sandlance off the sandbars in front of the house. Each morning we were greeted by the screams of young murre chicks which had temporarily misplaced their parents after a dive. This situation caused an interesting dialogue between the murres out on the water and the murres behind our fence. Our chicks called and the adult murres answered, to build up an intriguing dialogue.

While most of the birds were feeding on fish some way out from the shore, about fifteen hundred surf and white-winged scoters dug small molluscs and shellfish from the sandbars that stretched along the beach.

Sometimes we wonder why we roam so far afield when at our doorstep are some of the most fascinating regions on the whole continent. Five miles to the north lies Mandarte Island, one of the largest seabird rookeries in the Gulf Island region, where scientists are presently studying the gulls, cormorants, and guillemots. A few miles further north from Mandarte are the Ballingall Islets, where old dried whitened Rocky Mountain juniper trees sprout the artistic bulky stick nests of the double-crested cormorants. The first birds observed breeding here were seen in the early twenties. By 1939 the pioneering tree-nesting colony had increased to thirty-three nests. On my first visit to the Ballingall Islets in 1967, there were still exactly thirty-three nests artistically adorning the beautifully deformed limbs of the dead trees. Thousands of individual sticks had been expertly interwoven between existing branches to fix the bulky nests to the trees. One nest, complete with fishing lure, was actually reinforced with hundreds of feet of fishing line, probably scavenged from a hapless fisherman. As if this wasn't strong enough, several laps of line were wound around adjoining nests. With its sleek gray-and-black-mosaic plumage, its brilliant green eye, bright blue mouth, and contrasting orange gular pouch, the double-crested cormorant looks every bit a gay remnant of the prehistoric past.

Close to the city of Victoria are Chain and Trial islands, which shelter large nesting colonies of pelagic cormorants. City residents are fortunate that the slogan "Follow the Birds to Victoria" has validity. Yet the close proximity of these nesting colonies to people is sometimes unfortunate for the birds.

Although the Gulf Islands are built up with summer houses and permanent dwellings, they still contain about ninety-seven pairs of nesting bald eagles. This is undoubtedly fewer than existed fifty or a hundred years ago, but nearly a hundred pairs in fourteen hundred square miles is still an enviable number compared to the few that exist in the entire United States.

One of the most enjoyable features of living among the Gulf Islands is being able to live off the land. Japanese oysters are abundant in the shallow bays and bars of the islands. Crabs can be caught in crab pots or by merely picking them out of the eel grass. In some places one can gather abalone without having to dive. Salmon can be lured to a line as one stands at a sand spit, codfish jigged from any kelp bed, clams and mussels, sea urchins and sea cucumbers abound for the gourmet.

The Gulf Islands in Canada and the San Juan Islands in the United States are a continuous archipelago with the border running between. The more southerly San Juan Islands, in the rain shadow of the high Olympic Mountains, receive less rainfall. Consequently on these islands there are more open areas, thinner stands of fir, and a greater profusion of cactus. Seabirds are scarce, eagles are more so. Ninety-seven pairs of eagles nest in fourteen hundred square miles of the Gulf Islands, whereas only about five pairs nest in three thousand square miles of the adjacent San Juan Islands.

David was anxious to investigate the San Juan Islands for peregrine falcon eyries as well as to make a census of the alcids. With such a good excuse for a brief respite from construction work at Island View, we left Mother with the animals and invited Hilary on a boat trip.

Father lifted his eyebrows incredulously and made some caustic comments when he saw three people, sleeping bags, spare gas cans, food, fishing and camera gear piled high in our fourteen-foot aluminum dinghy. David assured him that unlike our northern trips to remote and isolated islands this voyage was to more civilized areas where we would seldom be far from people. The American islands are far more populated and settled than the Canadian ones.

First stop was Roche Harbor, a few miles to the northeast, where David as skipper of the boat and the only one to land climbed the dock to signify our entrance into the United States. Our second stop was Friday Harbor on San Juan Island, where we obtained a three-day cruising permit.

Washington's seabird rookeries are all classed as wildlife refuges and each displays a large notice prohibiting landing. Although we were armed beforehand with a fascinating number of special permits, I was most conscious of such notices as boats circled around to take pictures of us and float planes hovered overhead or swooped low as we made our brief intrusion to count numbers and species. On two of the low-slung rocky islets which had been used for target practice during the last war, the gulls and pigeon guillemots were nesting in the old bombs themselves. The sun burning on the metal made each nest an oven—a built-in baby sitter for the parent birds.

Despite brilliant sunshine the strong winds on the open water made the journey south to Whidbey and Deception Island a chilly one. We counted each puffin and murre on the water, each guillemot on the rocks and commented again on the smaller population. The records show Deception Island had once been the home of nesting rhinoceros auklets. We now wondered if Deception Island was the destination of the auklets that nightly fed at Island View.

We were deceived. The island provided just one small pebbly, highly shelved beach to land the boat. Helped by a nudge in the rear from Hilary I climbed to the top. The burrows so hopefully observed from the boat were deserted except for the homes of feral domestic pigeons. We circled the summit of the island: a systematic search of all burrows yielded nothing but a few guillemots. Perhaps there would be a night flight of rhinos, as on Solander and Triangle when the parent birds coming home to roost would reveal their own burrows. We set up the tent and waited. Night came still and dark with no whir of black wings to break the monotony. At midnight Hilary and I fell asleep, but David out in the boat continued to look and listen for rhinos.

Next morning we continued our survey along Rosario Strait up the east side of San Juan Island and back to Orcas Island. Falcons nest on high windswept bluffs. At each potential eyrie site David fired a pistol into the air to flush possible falcons, but no birds were observed.

Hilary hoped that David would stop for a while to allow

suntanning or beachcombing. I will be the first to admit that the Hancocks are incapable of sitting still. Poor Hilary! She had spent most of her weekends with a shovel, rake, cement trowel, or hammer in her hand. Now she was bumping her way past the scenic attractions of the San Juan Islands in a leaky boat with instructions to concentrate on birds and bailing, not basking and baking.

Three hundred and fifty miles later when we eventually returned home Father seemed surprised to see us.

Three days later we were catching the Port Angeles ferry. Dr. Allan, of the World Waterfowl Research and Propagation Center in Salt Lake City, Utah, had asked David to collect and raise for him several of the alcids, particularly the rhinoceros auklet. We were eager to observe a new area, little known to zoologists, so we chose the Washington coast for our expedition. Armed with nine permits from seven different government agencies we drove to the ferry.

"Yes, we're going down the Washington coast for rhinos!" was David's reply to the bewildered customs official. "Well, would you believe murres, tufted puffins, and pigeon guillemots?"

I tried to curb David's fun when I saw another customs man dart in and out of traffic and run madly up to the front of our truck. The newcomer's first question sounded alarming: "Well, what are they after this time, Fred?"

Our bewildered customs official sounded a little sarcastic as he replied, "They say they are after rhinos! I think they're giving me a hard time. Even I know rhinos are only found in Africa."

"It's O.K. Fred. They are just after *Cerorhinca monocerata*," he answered with a wink to us.

Now we were the ones taken aback. Who would be able to

quote the scientific name of a rhinoceros auklet, our weird-looking and seldom-seen quarry?

The stranger continued, "I recognized your beard and all that paraphernalia on your truck. I often watch you on television, Mr. Hancock. My, I'd love to get up to see some of those islands you visit in Canada."

To our amazement here was an ardent birdwatcher who made helpful suggestions as to where we might go to find rhinos.

Our destination was the many small islets of the Cape Flattery Rocks, Quillayute Needles and Copalis Rock national wildlife refuges, where several hundred thousand birds receive government protection.

From Port Angeles we drove into a superb sunset around the great ice peaks and lakes of the Olympic Mountains, through the vast rainforests of its slope along Route 99 to arrive after dark at La Push. Famed seasonally as the capital of silver and king salmon fishing, this busy fishing village and Indian reserve on the Washington coast was the embarkation point for our first target, Carroll Island. The two cafés were already closed, probably to be ready for the first of the dawn fishermen. We retraced our route along the road and climbed a side trail high above the village to pitch our tent on a wide, freshly bull-dozed field.

After a hard night on the ground we awoke next morning to find ourselves under a clammy damp blanket of sea fog in an Indian graveyard! Closer inspection of the graves revealed that the cemetery had been many years in operation. Plastic flowers decorated the deteriorating grave sites.

We checked in at Bud's Café for a breakfast of hotcakes and coffee, then reported to the Coast Guard station.

In preparation for the present trip I had read Karl Kenyon's scientific paper on the Washington seabird islands. His survey showed the islands as jagged, precipitous bare rocks which seemed even worse than Triangle.

Our conversation with the Coast Guard did nothing to alleviate my apprehensions.

"Where are you headed?" asked the handsome young coast guard with the crew cut.

"To Carroll Island and later to Destruction Island."

"Carroll Island, eh? Hell of a steep island!" he answered, looking straight at me with a look of surprise.

"Don't tell her that. I've been saying how flat it was!" came David's laughing reply. There were chuckles all around while I smiled weakly.

"Well, come looking if we don't report back in five days," David called back as we descended the steps of the immaculately painted red-and-white building.

Similar looks of surprise and incredulity followed us down at the fish dock as we loaded our fourteen-foot aluminum dinghy. Indians were setting out in their long dugout canoes, modernized with Mercury outboard motors. The gill netters were already out on the fishing grounds. The last of the sports-fishing tourists were setting out to join the throng offshore. Our boat settled lower and lower as David packed camping gear, food, water, rope, sacks, photographic and survival gear, and the well-ventilated and sandpapered crates for the birds. I threw over to him a green plastic sack filled with clothes.

"What have you got in there? It looks like enough for a month."

"No—only a swimsuit, a pair of shorts. Oh, I couldn't make up my mind about which boots to take so I brought the lot. That means three types of footgear."

There was general laughter from curious bystanders who

had interrupted their own fishing preparations to investigate this strange couple piled to the waterline with such unusual equipment. I could almost read their thoughts. Where would that girl be going? Why, there's only a bunch of rocks with birds on them and then the open Pacific. Well, they won't be going far in that dinghy!

We may have been overloaded but David always prepares for any eventuality. I must admit that lying flat on the tarp over all the gear was the most comfortable I've ever been.

No sooner were we out of the Quillayute River mouth than we were in the open Pacific. But today the sea was glossy smooth: only the gentle three-foot rise and swell remaining from the last storm indicated she could be anything but tranquil.

A few hundred yards from shore rose the Needles, several cone- and cake-shaped islands about three hundred feet high, bare up the sides, green on top. We skirted them to count the birds and make notes on their activities. Near Bald Island a flock of phalaropes flitted above the tide line. Many oyster-catchers wheeled by in groups but we observed no young on the intertidal rocks.

Suddenly a great gray whale emerged within twenty-five feet of our boat. The explosive puff of his exhaled air so startled me I nearly dropped the binoculars overboard. His blotched and barnacled scored back was almost within reach. Then he was gone—but not forgotten, for at that moment his hot oily, rancid expiration of breath settled around our boat. The gray whale is the only whale that feeds right into the shallows. It is distinguished from the other whales by the little hump on its back.

Near Jagged Island we were the ones to startle a huge Steller sea lion with her pup as they exploded to the surface beside our boat.

The water was alive. Herring boiled to the surface as they were pursued from below by the salmon. Regal murres swam by with their young chicks close behind. Groups of gaudy puffins sat on the water, fish dangling from their parrotlike beaks. Gulls screamed overhead and dived at the surfacing fish. The birds swam and bathed, seemingly oblivious of the hundreds of small fish boats in the near vicinity.

I had often dreamed of anchoring Triangle Island among the Gulf Islands so that more people could appreciate its beauty and its bird life. Yet I had always worried about the destruction and disturbance that human encroachment can cause. Now here were all my favorite species within three hours of Island View Beach and completely unconcerned about the boats and people surrounding them.

As we neared Carroll Island the birds' activity increased. Puffins popped to the surface all around us, their wings still beating from their underwater flying. Three of them gave us such inquisitive over-the-shoulder glances that they collided briefly, then reorganized their wing beats without losing a fish.

We circled the island. To my dismay, much of the rock surface was sheer wall but David, who had visited it before, knew of a route that gave surprisingly easy access to the top.

As usual the puffins, murres, and guillemots that fly to and from their young during daylight hours were readily seen. However, our aim was to study the nocturnal rhinoceros auklet so a systematic search of the burrows was necessary. One of the small burrows yielded a petrel. Quickly David returned it to the darkness of its home but not before it had regurgitated an oily mass of partially digested shrimp and shot it directly into his face.

While fighting through the head-high salmonberry David picked up three salamanders which he hoped would be a record for this island. Hilary! We thought of her simultaneously.

"She'll love these in her terrarium," David said. "Take them to the boat and wrap them in a safe place. Perhaps you had better stay down there to keep the boat off the barnacles while I search more burrows. I don't really think Carroll Island has many rhinos. I think there'll be a better colony on Destruction Island. We'll head there later."

I climbed down to the boat. Within an hour he had completed his survey of the island to conclude that rhinos would be too difficult to obtain on Carroll. To make the most of the calm weather he decided to retrace our steps to La Push, then continue on to Destruction the following day.

The return trip was wet and uncomfortable. David in the stern bore the brunt of the full force of the waves and was soon soaked to the skin. I had to sit on the tarp in the bow where I bumped, bumped, bumped till I thought my heart would never bounce back from my intestines. No wonder the rivets leak!

At last we reached La Push and pitched the tent beside the water. Over cheese and sausage we watched the fish boats returning to weigh their catches.

A night on hard ground made waking easy. The fog lifted a little as next morning we drove south to the Hoh River Indian Reservation. After following one false trail, which led us into a swamp where we spent the afternoon drying out our distributor, we eventually reached the mouth of the Hoh River and put the boat into the water. Here the Indians were netting fish in gill nets strung along the river to block the salmon on their upward migration. In this tidal estuary the salmon were silvery sleek, not yet the unpalatable black, green, and red of the imminently spawning fish.

In British Columbia the native Indian may catch salmon

in the rivers from the mouth to the highest lakes only for his own consumption, not for profit. However, in Washington the Indian not only may catch salmon anywhere in the river but he may sell them commercially through the marketing board.

The Hoh River faces into the open Pacific with large ocean swells constantly pounding onto the wide but shallow bar of gravel and sand which has been built up by the combined actions of wave and river. I looked apprehensively at the rollers breaking over the bar. The problem was to ride over the bar on the crest of the right wave, so there would be enough water for the motor to propel us to deeper water. Luckily for our gear, David was the navigator. Without mishap we safely crossed the surf line to a calm silvery-gray sea. Destruction Island was a shadow on the horizon six miles to the south.

Destruction Island may have lacked the majestic appearance of the Needles and its name may seem prophetic but it was a zoologist's wife's dream. As we threaded through the reef-studded shore to land in a delightful cove we trod not on barnacles, pebbles, or logs, but on soft yellow-white *sand* that reminded me of Australia. Nearly a hundred harbor seals were out on the adjacent rocks. Leading from the cove right to the summit was not a sheer wall but a *ladder*. There were a few missing rungs to negotiate, but how much better than dangling at the end of a rope or clinging with my fingernails to slipping shale.

And, even more fortunate, rhinos were in abundance. Honeycombing the hillside all the way to the top were thousands of burrows, all easily accessible under the stairway. And then waiting for us on the last rung of the ladder was an official greeting party to invite us in for a hot cup of coffee! I couldn't believe my luck.

"It's incredible. Civilization on a seabird trip," I whooped before David had a chance to introduce me to the Coast Guards.

Unlike the desolation associated with most bird bazaars, here was a large Coast Guard station with a substantial two-story red-and-white house, more than ample room for the three coast guards living there. Apparently our offshore movements had been followed in case we needed help.

We were surprised that the boys were unaware of the abundance of rhinos on the island. They accompanied us back to the ladder to watch us dig far under the roots of salmon-berry for the bizarre-looking rhinos. With little effort David grubbed out a few fluffy chicks. He caught one adult while it was cleaning out its burrow in readiness for the next breeding season. Its chick had swum seaward or else died, or perhaps the adult had not succeeded in hatching the egg.

When our limit was reached we wiped the dirt and perspiration from our faces and rested on the ladder before taking the remaining crates to the boat.

One of the fascinated coast guards, still holding his first rhino chick, had a good suggestion, "Why don't you come up to the station for a hot bath and home-cooked supper while we carry your gear to the boat?"

Surely an African safari complete with porters couldn't have equaled this superb ending to our Washington Coast rhino hunt! I looked at David. I couldn't believe this trip was happening to me. But, as I might have guessed, David replied, "Great idea, but we will have to take a rain check: it is best for the birds to get them back as soon as possible. If we rush we might make the last Port Angeles ferry to Victoria. Many thanks all the same."

As a result of our articles on the alcids in local papers the customs officials in Victoria knew the identity of the birds in our boxes and we had no delay in reaching home.

Next day I put the adults in the salt-water pens, keeping a constant watch on them to ensure that their waterproofing was adequate. As an experiment to emulate the wild situation as much as possible, I constructed long wooden burrows in which I placed a chick at one end and a plate of vitaminized herring strips at the other. In theory the chicks were to keep clean by walking to the end of the burrow to eliminate and to receive the proffered food: in practice I found that they remained huddled at the closed end of the burrow and seldom came to eat. Without visual contact with human beings they were less likely to lose their fear.

Next I tried cement blocks with two entrances, laid on the ground in a large enclosure. But with my bulk sitting over them in the middle of the pen they crowded in the corner. The more aggressive started to peck the more timid. After a few unsuccessful experiments I returned to my original method of placing each chick at my level in a cardboard box lined with crumpled paper and gravel. Each morning I burned the boxes and papers and replaced them with new ones kindly supplied in quantity by the local supermarket.

David wanted to film and observe a rhino's first reactions to the sea. It is generally thought that rhinos, like puffins, are abandoned in the burrows by their parents and later tumble down by themselves to the sea. David focused the camera as a young rhino, still with its halo of fluffy down, waddled unsteadily to the water's edge and started swimming.

He yelled, "Keep an eye on it so that it doesn't go out too far. We don't want to lose it."

Instinctively the chick paddled strongly outward from the beach. I dived in. The rhino dived too—out of sight.

I dog-paddled and breast-stroked until I could see where the chick emerged: invariably it was many hundreds of feet further out to the open sea and my swimming was no match. When he was a mere speck on my horizon I called to David on shore to launch the boat and bring the salmon net. Out came David swerving past me and roaring over to the rhino; five minutes or so later he had netted the chick and returned to me.

"Want a lift back? Hang on to the side of the boat," he shouted as he zoomed in close.

I could have swum the three hundred yards to shore but it was more fun to be dragged through the water. Father and Mother were waiting for us on the beach. A few others stood around looking worried.

"What were you doing that far out?" Mother asked. "One of the neighbors phoned to say some irresponsible child was in trouble in the water. Someone else thought you were a seal. Nobody in their right mind goes swimming that far out in the ocean."

After explanations and with the rhino safely back on the beach I suddenly realized I was cold: in the excitement of the moment, I hadn't noticed the water was freezing.

When the summer was over Mother and Father returned to their own home and David went back to the university in Vancouver to assemble his eagle data. I started teaching English at Monterey Elementary in Oak Bay, about fifteen miles from Island View Beach, a distance that demanded I learn to drive.

It was at this time we bought a dog. With our penchant for wild animals I hardly thought we needed a dog, but David had long been impressed with the Hungarian Vizsla, a pointing dog with a smooth coat, sleek lines, and friendly nature. We called our effusive floppy-eared pup Haida and hoped she'd like being a playmate for Sam.

Sam lived loose in the completed compound with his companion Haida. We didn't give him free access to the sea at our front door in case he swam out a little way, became curious as he had in the Santiam River, and attempted another life on his own in the sea. Although he was fat and sturdy, we doubted that he was any better adapted to an independent life than he had been on his journey south from the Pribilofs or in Oregon.

Sam at his present weight of 250 pounds is probably the largest fur seal raised in captivity. Zoos have a poor record for keeping fur seals. For many years the fifty or so animals collected annually and shipped to zoos died from numerous common ailments which dogs, cats, and humans could easily resist. Few zoo specimens have reached more than a few years of age. The record for keeping a fur seal alive in captivity—nine years—was set by San Diego Zoo in 1944. At the time of writing Sam has a few months to go before he reaches that mark.

For some unexplained reason captive fur seals seldom grow at the same rate as their free-roaming cousins. Although they do not have the same diet as in the wild, the captive ones have been amply fed. Even Sam seems to be slightly underweight in comparison with his wild relatives.

Sam's herring intake depends on the time of the year and the temperature of air and water. His responses to the temperature of his immediate environment are superimposed on his basic natural pattern of seasonal movement. In the wild state during the spring and summer as an immature bachelor bull he would be living, sleeping, and eating on the outskirts of the rookery, making a few raping forages after the females or test-ing an occasional uncertain harem master. Later as a harem master of ten years or more he would haul out on the rook-ery to establish territory in preparation for pupping and mat-ing, then spend the entire summer defending his harem. He wouldn't leave his territory to drink or to eat. Similarly Sam's appetite lessens in the summer months.

But another factor is involved. When Sam is hot he has little concern with eating, which is an energy-producing activity. He may voluntarily go without food for several weeks. In the winter when the temperature drops, Sam eats up to thirty-five pounds a day.

In part his weight deficiency is due to a lack of excess neck blubber, which is the food and water reservoir of the seasonally fasting fur seals. As Sam doesn't spend long months in con-stantly cold seas he has little need of such a warm insulating blubber layer. To keep pace with his winter appetite I had her-ring delivered by the ton. Invariably it was during David's ab-sence in Vancouver that Willy Egeland, our good-natured fisherman, would phone to say a ton of herring would be at my front door in ten minutes.

Somehow I had to inveigle a friend into holding the plastic bags while I shoveled in the herring from Willy's open fish boxes. The bags of slippery fish had then to be tied and stored in the freezer. As can be expected I had few volunteers.

Island View Beach was idyllic in the summer but in the winter presented problems. July and August brought bright sun; September to November brought constant rain. Gales up to eighty miles an hour lashed at the dike, whipping up the waves to enormous heights and hurling foam-crested breakers at the house. Overnight the beach was stripped of a mile-long layer of fine gray sand to become a beach of cobble stones and pebbles.

The sea was an army storming us with cannonade of logs and driftwood. The force of the water ripped out the filter box, tore apart fourteen hundred feet of metal pipe, and sent both box and pipe crashing to our front door in a jumble of broken screen and twisted metal.

Undaunted, David turned to our back door. He hired a bulldozer, a Gradeall, and a backhoe to dig a four-hundred-foot reservoir, a series of ditches and a well to pump in fresh water from a shorter distance at the back of our property to the pools and tanks. The sea line was buried again, this time deep in the gravel layer, which was more protected. In theory we now had two water systems, fresh water at the back door and salt sea water at the front.

When he wasn't keeping our house around us, David commuted between Victoria and Vancouver, preparing several half-hour television films for CBC's *Klahanie* program, writing articles on our summer field trips, researching effects of pesticides on wildlife, or drumming up support for a campaign to prevent dumping of untreated sewage into the sea.

After three years of exciting adventures, nearly fifty thousand feet of color film, and enough memories to fill an encyclopedia, David began to prepare for our first feature film, *Coast Safari*. The film was to be the candid portrayal of the trials and tribulations of a biologist and his wife studying the wildlife of the coast. Now David hoped to spend his summers in the wild, subsequent winters lecturing with the film in schools and theaters around the world.

We named our headquarters at Island View Beach the Wildlife Conservation Center.

David felt that new approaches to conservation problems were needed: the Center, he hoped, would help fill the void in the wildlife conservation field between what the government agencies and what other established conservation societies are presently accomplishing.

Many avenues of wildlife research are too narrow in scope to justify large expenditures of government money. Educating the public to the values of wildlife is a function often neglected by governments. Unfortunately the good work being done by the nonprofit societies is often limited by lack of financial support. By design these societies primarily appeal to the already converted—hunters, naturalists, or birdwatchers. This is how it should be. These conservation-oriented organizations play an important role by lobbying the government. However, the vast body of the general public seldom realizes the value of our great natural heritage, and fails to understand why, when, and how it should be concerned, and what, as individuals, they can do.

In short, then, the primary function of our Wildlife Conservation Center is to get more people involved with and concerned about conservation.

Wildlife conservation research is one field where the researchers have failed adequately to present their case or objectives. Wildlife researchers might take some advice from medical researchers, who readily communicate with the public through news releases to the papers, radio, and television. The professional medical organizations long ago realized that to get money and support for their projects they must make the public aware of their objectives, their successes, and their continued financial needs.

It has been thoroughly established by the medical and social scientists that there is a sound foundation in the old saying

"Man cannot live by bread alone." In the hustle and bustle of modern society there is an increasing need for mankind to be able to find places for relaxation and recreation: escape from the cement jungles becomes not a luxury but an absolute necessity.

That wildlife biologists have not sufficiently exploited the mass-communications media to present their cause and their objectives is most surprising in view of the fact that they are dealing with a topic of universal appeal—animals and the great outdoors.

So David's work at the Wildlife Conservation Center is concentrated in two main spheres, first, research into topics not adequately covered by other organizations and/or with significant conservation value; and second, interpretation of wildlife values to the general public through writing, exhibits, lectures, and motion-picture production.

Now that we had more facilities we acquired more animals; in addition to Sam, Haida, the seabirds, and the falcons, Island View now housed some rejected tortoises, Gomer, Himie, and Ludwig; an abandoned bear cub called Bubu; and a few orphaned cougars, Tom, Oola, Tammy, and Lara.

Sadly, we have been given many birds that have been shot. Sometimes we can do nothing except end their misery. On one occasion a magnificent bald eagle with a seven-foot wing span was found wandering on the beach. A bullet had shattered three-quarters of an inch of bone in its left wing high up on the shoulder and the wound was widely gangrenous. Amputation of the wing was essential, but because of the gangrene David held little hope for the eagle's life. He phoned several veterinarians but none would help in the amputation except grudgingly and at an astronomical cost.

We decided to do it alone. Meanwhile several pictures of the wounded eagle were taken for the local press to help educate the public against further shooting of these proud and majestic birds. In response to one of these news releases Inga Cordsen, once a veterinary assistant in Germany, phoned to say she would come the following day to help with the operation. David drove down to the drugstore to buy what seemed to my inexperienced eyes a whole hospital of drugs and medicines.

Next day Inga arrived. She was a strong-willed and forthright woman who, unlike me, didn't flinch at sawing through the thick bone of the eagle's shoulder. Having no mind or stomach for such a grisly operation I did little that was constructive except to talk gently to the eagle and hold its taped beak, trying to take the place of the deep anesthetic that David was loath to give. Little research has been done to determine what anesthetic works well on birds. Not wanting to chance killing the bird with an anesthetic, David decided to use none. Even injections of antibiotic had to be given with great care. Penicillin with procaine, commonly administered to humans and most mammals, is highly toxic to birds: one small dosage will kill them within a few minutes.

After the wing had been cut free of the body, the wound was sewn up, liberally covered with an antibiotic ointment, and bandaged. The operation took several hours but recovery took several months. Irritated by the bandaged wing stump, the eagle picked at it. David overcame this problem by taping a block of styrofoam to the beak as a muzzle. Recovery was incredible. Right from the start she ate well. We named her Baldy and held out a little more hope for her permanent recovery.

All creatures require a territory in which they feel safe. At first Baldy was allowed to run and jump where she chose, but there was always some danger of the wound's reopening. As soon as she selected a favorite stump in a particular enclosure in the compound, that became her home. Now, several years later, she announces her territory in a loud and seemingly triumphant kack kack kack whenever anyone approaches.

A year after the operation David gave her company and an opportunity for experience in the form of a nestling eaglet, which she carefully raised.

Since then we have received other eagles needing amputation. Unfortunately these have all been immatures, so it will be some time before two amputees can be placed together for possible reproduction. Though it is commonly supposed that eagles mate in the air, David has observed otherwise: he believes that after an aerial courtship eagles mate on branches or the ground. Perhaps old Baldy can still fulfill her maternal function.

Both bald and golden eagles are protected by all provinces and states in Canada and the United States. In Canada shooting an eagle invokes a five-hundred-dollar fine; in the United States it is a Federal offense to shoot an eagle or remove their eggs or nests. There are loopholes in both laws, unfortunately. In Canada, if a person can prove he is protecting his property from wanton destruction by a predator he is free to shoot it. In the United States, if a governor declares the eagle to be an agricultural menace it may be poisoned or shot. Laws are good in theory but open to much abuse when the philosophy of most people is to regard predators as vermin.

I once received a newspaper clipping showing a lady hunter proudly displaying a shotgun in one hand and a dead bald eagle in the other. The caption labeled her "a heroine" and complimented her on her "bravery." I sent this clipping to David

Sam investigating the bald eaglets

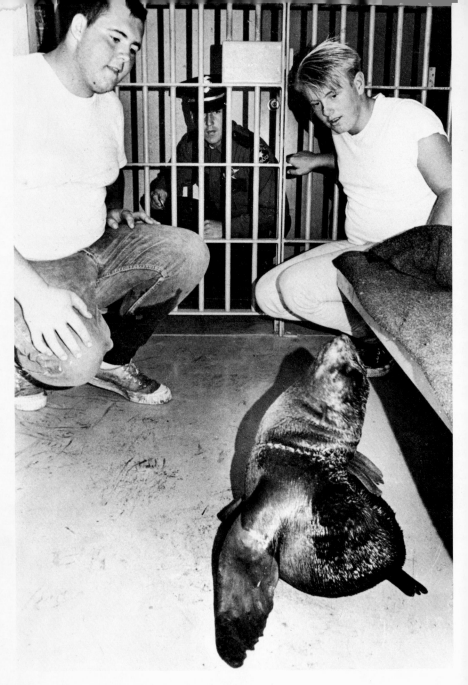

Sam in jail in Independence, Oregon (John Ericksen)

Facing page:
Sam: supercilious, alert, mischievous

Sam preventing Haida from leaving the yard

Sam cuddling up with his cat on the stove door

Peregrine falcon at nest

Four-week-old peregrine falcon from research breeding project being weighed at school

Facing page:
David uses a young peregrine falcon to illustrate a conservation lesson

Sam looking on greedily as Lyn feeds the eaglets

Facing page: A bald eagle in flight

David and Lyn band a bald eagle

Measuring an adult bald eagle

Facing page:
The catch: an immature bald eagle flies off with supper in his talons

On following pages:
Views of bald eagles in flight

while he was in Wisconsin attending a conference on the endangered birds of prey.

Recently a friend sent me a clipping from the *Cape Breton Post*, Nova Scotia.

THE EAGLE CHASE

An eagle was captured by town police Monday morning and his appearance created wide interest from local residents and visitors to town. Complaints were received that the eagle was frightening children and chasing dogs and cats and baby carriages. Police Chief Bernie Kelly and Officer Dan W. McInnis caught the eagle on Cottage Avenue. He was shot and viewed by many as they passed the Town Hall. The eagle weighed about twenty-five pounds and measured seven feet from tip to tip. His capture was something unusual for town police.

No eagle ever weighed twenty-five pounds: the record is sixteen pounds and the average is five to twelve pounds.

In a recent U.S. study 119 bald eagles found dead were examined to determine the cause of death: 90 percent had been shot. In the southwest United States many sheep and cattle ranchers still shoot and poison eagles, believing they kill lambs and calves. Research has shown this to be untrue. In Alaska, before it became illegal, ranchers used to shoot down hundreds of eagles each year from aircraft. The Alaskan Government once paid bounty on over a hundred thousand eagles—many times the number of eagles alive today—because it was thought eagles were harmful to spawning salmon.

Our first Christmas at Island View was a disaster. Victoria has the reputation of being the one city in Canada where roses bloom and all is green at Christmas, but that year of 1968, for the first time since 1880, the mercury dipped to below zero. Australians had record heat; Californians had record rain. And

Victorians had record cold. For weeks we were inundated under a white blanket of snow and of course everyone was totally unprepared. Even here by the salt water at the beach the driftwood logs were thickly encrusted with ice and to walk out the front door meant plunging into deep snow drifts.

Ironically—if that's the word I want—on that first night when the temperature plunged to zero the propane furnace ran out of fuel. Beach bark and driftwood for the pot-bellied stove were wet and covered with ice. There were no Presto logs or other wood available in the entire city. The thermometer registered six degrees in the living room and a hot-water bottle froze in bed. The refrigerator was the warmest place inside the house and the sea the warmest place outside. Water dripped from the cliffs along the beach and steam rose from the waves.

We devised our own central heating and carried it around with us. I wore two spencers (Australian item of thermal underwear), four sweaters, and a parka. For bed I removed the parka.

All the overhead water systems which ran across the ceiling and down through the walls froze solid, as well as the pumps in the porch inside the house. Seven days later there was a temporary thaw. Immediately the pipes burst and all the water trickled through the cracks and dripped from the ceiling. The pump split open and lay on the floor. After the thaw it froze again.

We laughed about living primitively so close to a city. But we couldn't laugh about the plight of the seabirds, which required constant running water for their very survival. Under normal conditions of ice and snow they would have no problem. But one night at midnight an assistant with the best intentions broke the ice that had formed over the pool. Immediately six of the birds dived under the ice and were drowned. To keep the seabirds clean was to keep them alive. We removed them from the large walk-in aviary and brought them all into the house.

The stench of bird excreta and herring was soon overpowering. With our electric stove the only remaining source of heat in our uninsulated house, the boxes of herring had to be thawed in and on the oven door. Kitchen and living room were piled high with herring in various stages of thawing. I remembered another time in the first year of our married life when a kitchenful of herring had sent me running down the street in horror: now necessity conquered aversion.

Holidays over, David departed on a month's lecture tour in northern British Columbia, where the temperatures were forty below zero. I started the new school term at Monterey and looked after the birds.

In a slight thaw the birds were returned to the compound and the house cleaned; two days later the mercury plummeted again to zero. I awoke at 6 A.M. to break up the frozen fish left there from the night before, thaw water and fish on the electric stove, leave out the newly thawed fish for the birds and return to the house to dress. It was heartbreaking to know that half an hour later the thawed fish would be frozen again before the birds had been stimulated to eat. Luckily a close friend, Elton Anderson, came at noon to thaw and replace the fish I had left out in the morning. On my return from school at 6 P.M. I melted ice, thawed fish, and fed the birds again, and the final thawing and feeding were at midnight.

By the end of February temperatures moderated and the snow rapidly melted. Then the birds had to be cleaned and the long process of re-waterproofing their feathers commenced.

At last it was spring. The cliff swallows returned to rebuild their nests on the wall above my newly painted back door. A pair of violet-green swallows built a nest in a new spot above our bedroom window. Several pair of barn swallows stood guard on old nests on windows behind the cabins.

Then the first of the starlings moved in. They started to drop dry grass through one of the knotholes in the cedar-facing of the freezer room. In three days the two busy parents had dropped a pile of grass four feet high, three feet long and a foot wide, all of which landed on the ground beside the freezer. Feeling sorry for their vain efforts, David built a ledge on the other side of the wall about six inches below the knothole so that their grass didn't drop into nothingness. Three healthy broods of the ubiquitous starlings were subsequently raised.

With the spring also came Sam's instinctive romantic urges to herd anyone and anything into his harem by his standard bluffing charge. His wide-open mouth and lunges from a con-certina-like neck are usually sufficient to keep all potential harem material in one corner.

At Island View Beach Sam didn't have Ginger the cat to hold on the oven door or the schoolchildren to chase between the alley and the street. He now enjoyed the freedom of our large compound. Here he was Top Seal, courting puffins, ducks, tortoises, and geese, but his main attentions were di-rected to our young Vizsla pup Haida. The Vizsla breed is known to be very affectionate, but Sam was too much even for Haida. For about half an hour the dog enjoyed Sam's over-tures and they frolicked together until Haida was bored or had her ear pulled too often to the tune of Sam's mating call. As soon as Haida edged toward the gate Sam, the ardent lover, grabbed her by the scruff of the neck and firmly propelled her back into the compound.

We are very anxious to acquire a female fur seal for Sam. Apart from waiting for one to wash up on the inside coastal waters as had happened with Sam, there are three other possibilities.

We might get one from the United States Fish and Wildlife Service but much red tape is involved: also, it would be a wild female, subject to the landborne diseases that had bothered Sam in his first year with us.

It would be possible to get one from the Federal Fisheries Biological Station in Nanaimo, which is responsible for killing fur seals on the open seas for analysis, but the animal would probably be wounded.

The Point Defiance Aquarium in Tacoma is a more hopeful source because the animal would be better adapted to captivity. Sometimes fur seals are found in rivers entering Puget Sound after straying from their annual migration route to and from the Pribilofs. So strong is their instinct to swim south that once inside the sound some are apparently unable to orient themselves north to get out of it. Instead they continue their journey southward up the rivers. When these storm-driven or lost waifs are found in Puget Sound they are brought to the Port Defiance Aquarium in Tacoma, where they are kept in the excellent care of Cecil Brousseau. When those maintained in the aquarium tank grow too large for their facilities, one of the females will be offered to Sam.

At the time of writing Sam is eight years old. In the wild there would still remain a year or two before he would compete with other bulls for territory on the rookery. So we—and Sam—wait.

9

Killer Whales and Eagles

DAVID HAD LONG CHERISHED A DREAM TO CIRCUMNAVIGATE BOTH North and South America in a rubber boat. We now acquired a sturdy fifteen-foot raft called a Canova and two brand-new Mercury outboard motors. David planned to test them in the stormy seas of the Queen Charlotte Islands, one of the most challenging waters of the continent.

Hilary has long cherished *her* dream to visit the Queen Charlottes, the ancestral home of the Haida Indians. I was teaching school during the month of the intended trip, so it was an ideal opportunity for Hilary to accompany David. Their expedition was the subject of "Haida Country," the first hour of our film *Coast Safari*.

Apart from an exciting eighteen-inch rip in the bottom of the boat, which Hilary sewed up with needle and thread, she and David returned from the islands full of enthusiasm for rubber boats. Holes seldom occur, but when they do they can be patched, and the boat was light enough to be hauled from the water when a storm arose.

We have now been riding on rubber for several years.

Many people still look aghast at our audacity in tackling the open seas in such frail craft. David looks at it another way. When disaster occurs in seas that are too rough to keep other boats afloat, it is our kind of rubber boat that is used in the official rescue attempts. So he reasons that we may as well start off in a boat in which other people find themselves finishing up.

It may be the safest boat afloat but it isn't the driest or the most comfortable, as I was to discover on our next expedition. Fascinated with the British Columbia coast, David planned to travel by rubber boat through the Inside Passage between Vancouver Island and the mainland, cross Queen Charlotte Sound, and thread past the inlets to Bella Bella. We had covered this route before by float plane during eagle surveys but now we wanted to film as well as observe the life of the coast more intimately.

Our destination, Bella Bella, was six hundred miles to the north. While many people nowadays are talking about conservation, some are actually doing something about it; for instance a young man named Sunny Dickson walks onto the wharf at Bella Bella twice a day to feed a group of about sixty eagles with meat and meat scraps from the local B.C. Packers' store. The significance isn't in the feeding but in the influence this action has had on the local villagers. Earlier in the spring when David had called at Bella Bella on his way to the Queen Charlottes, he had asked the storekeeper if there were many eagles around. Knowing that eagles often get shot around fishing camps, David was pleasantly surprised when the storekeeper abruptly told him: "You'd better not shoot eagles around here. We feed them and we like them."

David assured him that he had no intention of harming a feather of any eagle, and he was interested to learn how it had come about that the eagles were so strongly protected in this community. He learned that since Sunny began to feed them, and the people of Bella Bella got a chance to observe the flying skills of the eagles, they began admiring the bird for its true magnificence and, admiring it, they wish to protect it. David wanted to record Sunny's influence on film in *Coast Safari*, because it illustrated his own philosophy of conservation, that it is important to show that the wild has a meaning and value to each of us before we can expect each individual to respect that wild.

Our expedition was launched from our front door. A small group of incredulous picnickers gathered on Island View Beach to wave goodbye as we headed into the Gulf Islands.

Thousands of city dwellers from all over the Pacific Northwest flock to this island playground for its picturesque scenery, mild year-round climate, excellent fishing, and leisurely way of life. Many of the smaller islands are privately owned by the richer Americans who have flocked to Canada to buy some of the sanity that has been lost in the crowded cities to the south.

Luckily there are still some islands that belong to the Crown. Portland Island is one of these. I was rather amused when David told me how this island became public property. Evidently the Parks Branch of the Provincial Government bought the island as a gift for Princess Margaret when she visited British Columbia as part of the Centennial celebrations of 1958. It was expected that she would return it to the people of the province as a marine park. Several years passed and nothing more was heard. The Parks Branch personnel were wanting to get on with the Park's dedication but still it remained the personal property of Princess Margaret. Finally a very embarrassed provincial secretary traveled to England to ask Her Royal Highness if she would mind returning the gift, in return for which the province would call it Princess Margaret Marine Park.

The Gulf Islands' reputation of sunny skies and warm temperatures fell flat as we edged along Saltspring Island. Blue skies clouded black and the subsequent downpour gave us our first chance to test our new yellow rain slickers. I hoped we were going to have all our troubles in one day, because at the same time a pin broke on the steering arm of the outboard motor. Luckily the patrol boat for the Royal Canadian Mounted Police was fishing nearby and the officers cheerfully towed us into the village of Ganges on Saltspring Island.

A few hours later the arm was repaired and we crossed Trincomali Channel to long finger-like Galiano Island, where we slept in the shed of an old abandoned orchard. Despite the constant rain of this particular summer we found no reason to bother erecting our new tent. In many places along the B.C. coast there is always some shelter: a freight shed, floating fish camp, a linesman's cabin, or an old shack that still stands as mute testimony to the pioneers who once settled this rugged inhospitable shoreline.

Next day it was still raining as we continued along Galiano and Valdes islands. Our box seat for birdwatching on Galiano was provided by scraping glaciers and pounding seas that created delicate patterns in the sandstone thousands of years ago when the Gulf Islands first reared from the deep. Now we could tie our boat to a shelf hollowed out in the sandstone and sit a few inches above the sea to stare up at eroded holes that were ready-made nests for cliff swallows, cormorants, turkey vultures, and pigeon guillemots.

We walked the shorelines of many of the islands to examine the ancient shell middens or garbage dumps of the Indians.

Strolling through the dilapidated remains of some deserted buildings, I came to one very low structure that had a recently caved-in roof. Supporting myself on some of the fallen timbers and driftwood, I leaned over to peer inside. At that point David came up behind me to startle me with "What are you doing holding on to a human leg?"

I thought he was joking until he lifted up a rotting plank and there staring up at us was the mummified corpse of an Indian woman, dressed in mildewed cotton and lace and with her arms folded demurely at her waist. A gold tooth shone from her open mouth. We were amazed that in this damp climate the body was so perfectly dried and preserved. There was only one logical explanation: we had stumbled on an Indian grave that had very recently been vandalized. It was now our obligation to mend the tomb as well as we were able.

The day's continual downpour proved more than a match for our rain slickers. Wind slashed our faces with sharp needles and water soaked in at wrists, waist, neck, and ankles. By evening we had one thought—to reach Jackscrew Island. I remembered that Hilary had once sheltered on Jackscrew Island in front of a huge open fireplace in a palatial abandoned log cabin that featured moose jaws and deer antlers hanging from the walls and a grand piano in one corner. Its American millionaire owner had died and the lawyers were not able to trace the next of kin. That had been three years before, so we didn't hold much hope that it was still empty. However, in these more populated parts of the coast, I didn't feel as much like pioneering as I usually do. I fully intended to ask the present owners if they would mind two rain-soaked rats in a rubber boat sharing their fireplace.

Arriving at last at Jackscrew Island we climbed the steps of a sturdily constructed cement jetty and, seeing a flickering light through the trees, found our way to a large log cabin with a wide stone verandah. The great oak door hung sideways from its hinges: the new owner seemed anxious to retain an aura of elegant dilapidation.

We stepped over the threshold into a huge room which reminded me strongly of a medieval castle. The moose jaws and the deer antlers had been ripped from the walls, the roof, windows, and doors were battered and broken, but the grand piano was still there as Hilary had described it. Unfortunately the intricately carved legs had been sawed off and the grand old lady was now reduced to her knees.

But what appealed to us most at that moment was a roaring fire blazing in the open hearth, and to welcome us, another couple of soaked mariners who had taken refuge from the rain. We shared their lines to string up our wet clothes in front of the flames and changed into dry things. A huge oak dining table ran almost the length of the room, and after a meal of oysters and ling cod we all bunked down in our sleeping bags in front of the dying embers in the fireplace.

The next day we left the Gulf Islands at Gabriola Island and continued northwest along a coastline of sandy beaches in waters protected by the islands of Georgia Strait.

Quadra Island off Campbell River guards the northern entrance of Georgia Strait. Here lies Seymour Narrows, the narrowest passage of water between Vancouver Island and the mainland, and the one-time home of the famed Ripple Rock, a shipping hazard that was eventually blasted out in the largest non-atomic explosion ever performed.

David had timed our arrival to shoot the rapids at peak flood, not only to test the boat in rough water but to record on film the excitement of ripples, rapids, and whirlpools. The man who filled our gas tanks at Campbell River warned us to avoid the turbulent water at the change of tides and graphically described disasters that had happened to boats that dared that particular waterway.

But David was full of cautious optimism as we sped toward the Hole in the Wall at Seymour Narrows. Above the noise of the water he shouted, "We have company. Look over there!"

Beside us through the single narrow channel between the whirlpools swam a pod of a dozen killer whales.

I was terrified. When the five-foot-high dorsal fin of a big bull cleaved the water beside our rubber boat, I wondered how we'd fare swimming in the whirlpools if that twenty-six-foot whale surfaced underneath our fifteen-foot craft. But David, in such circumstances, always has a quip to relieve the tension.

Killer whales were labeled "voracious killers of the deep," and much mystery and misunderstanding surrounded them until in 1964 the first was captured. Although there were no authentic records of their attacking man, they were known to eat seals, sea lions, porpoises, various fish, squid, and seabirds.

Dr. Murray Newman, Curator of the Vancouver Public Aquarium, sent a party to harpoon a killer whale, in order to prepare an exhibition. They caught but did not kill a fifteen-foot killer whale, which they named Moby Doll. Scientists flocked to Vancouver to study Moby Doll's physiology, behavior and voice, but the whale died in captivity after three months.

Killer whales may be sighted anywhere along the British Columbia coast, or even in the open water, sometimes far from land. They are particularly common around the Gulf Islands, the San Juan Islands, and Puget Sound.

As the pod continued to spout on either side of our boat I remembered David's description of the way the killer whales had used wolf-pack tactics to trap and eat the minke whale in Barkley Sound. Apprehensively I watched and wondered where the next whale would appear.

Suddenly the big male surfaced out of the sea like a submarine. Clearing the water in a vertical leap, its gigantic chest shining white in the sun, it stood up on its tail to examine us more closely. What tiny eyes for such a gigantic mammal! A flashing glimpse of his jaws and a dozen pair of long conical white interlocking teeth didn't make it easy to remind myself that in captivity killer whales have been found to be intelligent, docile, and friendly.

I wasn't only worried about the whales. Pieces of driftwood still sprouting branches were spinning crazily inside the ever-changing whirlpools. Just off Maude Island an enormous log was sucked like paper into the center of a gigantic gaping hole in the sea. The outer circle of swirling water, which gobbled down logs the size of our boat, had a diameter the width of a house. As we edged toward the spout, the speed of the churning water increased. Although the vortex was a mere six feet across at the top, it descended in a vertical column much deeper than we cared to get close enough to see.

As if he read my thoughts, David commented, "I hope the motor doesn't stop. We may be only making one mile per hour against the tide but at least we are on top of the water rather than down one of those holes."

Filming was disappointing. With me concentrating on avoiding the whales and David on avoiding the whirlpools, it was very difficult to capture the excitement of the moment on film. The camera should have been on another boat looking across at us. Neither of us wanted to cut the motor in case we fared no better than the driftwood did in the whirlpools.

David didn't think the Seymour Narrows was sufficiently exciting for film so he decided to try the other rapids near Stuart Island at the head of Bute Inlet. At the Arran Rapids the waters raged against us with such force that we were completely at a standstill. We were forced to detour around Stuart Island and try the Yucultan Rapids. Even staying as close as possible to the shore we made less than a mile an hour. As rushing water on the screen never looks as exciting as rushing water in reality, we gave up and drove the long detour around Stuart Island. At this favorite cruiser anchorage and fishing resort for American tourists, we stopped long enough for gas and clam chowder. My daydream of sitting in front of an open fire in a cedar-and-glass A-frame cabin was shattered with a typical suggestion from David: "We may as well camp on that bare rock above the tide line in the midst of those swirling waters if I can fix up an anchor line. It'll give us a grandstand view of the hundreds of eagles I've counted today along the shores of Stuart Island. We'll see if we can catch any eagles for banding."

After painstakingly anchoring the Canova to the rocks in the fast-moving waters, David suggested I strip down to a swimsuit and dive for starfish, which were clearly etched against a sandy bottom twenty feet down. The film was the only thing that persuaded me to make an ungainly splash into that frigid forty-degree water, so I was quite annoyed when the camera ran out of film at the beginning of my dive. While David changed the rolls I numbly clung to the Canova trying to keep at least part of my anatomy out of the icy water till the cameras were again ready for action.

While David caught some ling cod to bait his eagle traps I set up the tent for the only time we were to use it that summer. We spent the following day watching twenty eagles wheeling and diving for hake, which kept boiling to the surface in the eddying waters. I was curious about why so many hake were dying.

"Probably too much oxygen in these constantly moving waters," David replied. "Now you steer the boat away from the whirlpools and into the best position for me to film the eagles as they swoop down for hake."

As usual, it was easier said than done. Tempers were a little strained and I had vowed to throw myself into the rapids by the time we had the necessary sequences of film. With all that food available for scavenging the eagles would not descend to our ling cod, so David abandoned the idea of catching any.

Steep-sided mountains closed in on us as we cruised along narrow Johnstone Strait past Hardwicke Island to the mainland inlet of Port Neville. This is the home of the Hansen family, who settled here in 1895 and who now eke out a pioneer existence fishing, beachcombing, and running the government post office and gas station for summer fishermen and boaters. All that remains of a former townsite is the log-cabin store and post office built in 1925. David adores poking around in junk shops so he was in his element investigating the old store, which still contained supplies from the turn of the century—old sewing machines, Victrolas, books, toys, calendars, and even canned foods. At times like these I'm glad we travel in a rubber boat. It does hold down David's impulse to collect antiques and discarded things that he "might just use one day." In a meadow behind the old schoolhouse was another collection of rusted farm machinery, hulks of dilapidated boats, remnants of a barn, and an overgrown cemetery.

The Hansen family offered us a bed in their net loft and a salmon for breakfast. We were awakened at the first glimmer of dawn by a pair of chittering swallows flying around the room. They were nesting above the window I had closed in the dark before going to sleep. As a mink had stolen our salmon during the night we made do with a breakfast of bacon and eggs cooked on the Coleman in the freight shed.

Northwestward from Port Neville our progress was impeded by a twenty-five-mile wind, fog, and driving rain. All day we struggled slowly along the strait.

At Hansen Island, where we stopped to study a raft of several hundred rhinoceros auklets, I changed places with David at the motor and headed for what I hoped was the fishing community of Alert Bay on Cormorant Island. As each wave drenched my rain slicker I realized for the first time the wet and vulnerable position of the skipper at the stern. Soaked to the skin and with eyes so stung by salt that I couldn't see, I made a guess at where the dock might be. I knew we had arrived when our rubber boat bounced harmlessly off some heavy obstacle and David jumped up from the bow to wrest the motor from my frozen grip and shove it into reverse. I had hit the *Laurier*, the Fisheries Patrol boat tied alongside the dock at Alert Bay. As I struggled out of my rain gear I heard someone shout, "My God, he's a woman!"

I thought that must be evident by my doubtful prowess as a navigator. We were invited aboard for the usual welcoming cup of coffee. Some of the crew had accompanied us on the Triangle Island trip; those who hadn't had certainly heard of us and there seemed to be a lot of chuckling and stares. With the same air of friendly amusement, Captain David Waters helped me over the railings and offered us the dining-room floor for the night.

Alert Bay was once a small saltery to preserve salmon prior to shipment to Victoria. Now it is a fishing community with a large Indian population. Most of its buildings are clustered along the one main road which skirts the sea. It is the center of the Kwagwlth people, a competitive and progressive group who have built a full-scale replica of a primitive community house in which they perform Indian dances each week throughout the summer. Inside the longhouse tourists marvel at the mighty hand-adzed five-ton roof beams that had been raised in the manner of the Indians of a century ago. Another favorite spot is the cemetery. Each grave is marked by totem poles, all looking out to sea in the traditional manner.

From Alert Bay it is a short distance across to Sointula on Malcolm Island. This time David was at the helm. There was no wind and the sea was calm, the sky overcast. As we approached the dock I released the bowline, stood up on the bow, stretched my leg over to the dock and jumped—and missed the dock.

Water quickly saturated my woolly cap, down jacket, sweaters, jeans, thermal underwear, and heavy hiking boots, and I found myself sinking slowly under the weight. Instead of grabbing the camera, my gallant husband grabbed me and pulled me up, sodden and heavily waterlogged, then proceeded to squeeze me out. Some summer boaters looked on and laughed. On the dock above the float David found a trolley-type freight cart which he wheeled beside a wall to provide a certain amount of decency, and I changed into dry clothes.

Despite its oil-slicked wharf I like Sointula. Its atmosphere is unique on the coast. It was founded in 1901 by a group of Finnish settlers dreaming of a Utopian community. The original colony failed, but many Finns remained to form the core of the present village of eight hundred. Each house, spotlessly clean and well painted, is widely separated from its neighbor. No garbage or junk is thrown off the floats to the sea burial common in other communities. The friendliness and charm of the Sointula people is typified by the caretaker of the Finnish

Community Hall who, when I visited Sointula later to show our films, met me at the door and threw her arms around me in a gesture of enthusiastic welcome.

North of Sointula by six miles we ran into a most unusual procession. Two small gill netters were trying to tow a seine boat broadside through the water. Behind the seine boat in a floating net pen could be seen the dorsal fin of a young killer whale. David quickly tied up alongside the seiner and we climbed aboard. A cup of coffee was immediately proffered by burly dimpled Delmar Laughlin, skipper of the *Rick Son*.

"How did you get yourself a killer whale?" David asked, as he waved to the skippers of the accompanying boats *Barrie Lynn* and *Scallywag*.

"Well, it wasn't really me," said Delmar. "Yesterday Aaro Palo of the *Barrie Lynn* ran his gill net around nine of them. They were rolling and rubbing their bellies on the smooth boulders in the shallow water of one of the beaches on Malcolm Island. Aaro knew that several killer whales had been trapped at Pender Harbor earlier in the spring so perhaps there would be a market for more. Anyway it had to be an instant decision, so he made up his mind to encircle the pod."

Stephen Miller, the blond boyish deckhand, grinned widely. "Aaro corraled the whole pod in a few minutes, three big bulls, four small females or young males and two small calves."

"A salmon gill net held on the surface by a few corks encircling nine of the largest mammals in the sea sounds a bit incredible," David prodded. "If the wind had picked up, your boat and net would have been blown ashore. And if you tried to tow the whales anywhere, they'd swim out under the webbing as soon as you hit deeper water."

"That's where the *Rick Son* came in," Delmar explained.

"Aaro radioed me for help. I collected Mrs. Palo from Sointula and we cruised out to help. We ran our heavy seine net around them. Boy oh boy, you should have seen them panic when we tried to purse up! Barrie sure got a fright when the big bull panicked. He was in the six-foot skiff fixing more floats. In one charge the bull went right through the web, then the two other bulls tried the same but they got caught part way through. This made the net sink. While all the males were wriggling free, all but one of the remaining whales swam out over the cork lines the bulls had pulled under. As soon as poor Mrs. Palo saw that churning mass of whales head for the sunken cork line and Barrie right beside them in the dinghy, she screamed and I had to stop her from diving overboard to save him."

"Well, I notice Barrie and his dinghy are still floating. But I guess there are a few holes to mend," I broke in.

"I'll say—but it could have been worse," Delmar continued. "There's just Baby Doll left. Now perhaps if we can sell her we can cut a few of our losses. The tide will change soon and we'll start moving. We've been towing slowly for about sixteen hours. Probably get into Sointula in about seven hours. You're welcome to come along for the ride."

David's eyes lit up with the anticipation of new discovery. He wanted to study the behavior of the killer whale at close quarters. He knew that these animals were extremely sensitive creatures seeking company with their kind or men when in captivity. What family ties were being broken now that Baby Doll was separated from the other whales? Were the nine whales one family group? Was the large bull the sire and ruler of those with him?

Although more than twenty killer whales had been netted in local waters there was still much to learn.

The boats were hardly moving, waiting for the tides to

change. David and I spent the time trying to make friends with the whale. When David rubbed her head every time she spouted within touching distance, Delmar and Aaro were amazed at how soon the whale could be approached and how close we could come to her.

The family responsibility of a killer whale has been found to be remarkable. The family of another captured killer whale, Namu, followed him for three hundred miles before finally abandoning him to his captors.

All through the long tow to Sointula the eight free whales followed their captured comrade, constantly circling the procession of boats. Once, when Barrie Palo was again attaching more floats to keep the nets from sinking, one of the whales came up and carefully nudged the netting only three feet from his skiff. The whistles, squeaks, and screams of Baby Doll resounded off our dinghy, which was supporting one end of the floating pen.

These high-pitched sounds increased in number as the other whales called back. It was touching: each time we motored back from the tow boat to talk with the whale the herd showed more signs of concern. Often one or more of the accompanying pod would stand erect in the water, eyes peering above the surface to see more clearly what was going on.

When our procession was a mile out from Sointula, David spotted another pod of eleven whales approaching a mile to the north. Dorsal fins cleaving the air, this second pod resembled an army of knives bearing down in one straight line as if ready for attack.

The guardian whales were all countable and slightly to our south. The biggest bull, easily distinguished from the others by a tremendously high but bent dorsal fin, began a series of tail slaps on the water, making sounds like rifle shots.

Immediately the other pod stopped advancing. Some animals started tail splashing, others jumped clear of the water, but otherwise each pod kept its distance one mile apart.

By dusk our little entourage, the *Rick Son, Scallywag, Barrie Lynn*, and, of course, Baby Doll, entered the breakwater at Sointula, followed immediately behind by the family of the captive whale. By the congestion of fish boats in the colorful but crammed harbor it was evident that fishing was finished for the weekend.

The weekly shutdown gave Aaro a breathing space for negotiation and decision. He wondered how much valuable fishing time was to be spent in finding a buyer for the whale. Initial inquiries at Vancouver Public Aquarium and Marineland of the Pacific indicated there was little further market in North America for another killer whale, even a small female.

Meanwhile Baby Doll didn't lack visitors. All Sointula converged on the temporary pen that Aaro and Delmar arranged at the end of the wharf between their two fish boats. For a few nights David and I slept aboard the *Rick Son* and baby-sat a killer whale. All night long her whistles and high-pitched squeaks were answered by the vigilant pod out in the straits. We sat on the dock and tried to talk to her ourselves, hoping that our soft tones and company might be some small replacement for what she had lost. During the day we cut up salmon and tried to tempt her to eat, although David knew it was too soon for any success. But she did enjoy a back-scrubbing with an oar and broom. David and I were fascinated by this huge intelligent mammal, maligned so long through man's ignorance.

Our schedule called us northward. We wished good luck to the Finns and headed for Bull Harbor at the northern end of Vancouver Island.

The air was raw and cold, the wind strong, as we entered

Queen Charlotte Strait. Despite the swells over the bows our rain gear kept us comparatively dry until we sighted the *Skeena Prince*. Many and varied forms of transport are found along the Inside Passage between Vancouver Island and the mainland— kayaks, canoes, dinghies, fish boats, cabin cruisers, yachts, houseboats, and large bulk carriers. Busy tugs pull modern self-dumping barges piled high with logs or pyramids of sawdust for southern mills. Float planes buzz overhead. But the coastal freighters like the *Skeena Prince* are the lifeline of every fishing and logging community along the coast, delivering everything from bread and milk for the housewife to twenty-five-ton trucks for the logging camps.

Trying to manipulate both outboard motor and movie camera at the same time in the heavy seas, David crossed and recrossed the freighter's path. I felt sure we would be mowed down by the advancing hull of the *Skeena Prince* that loomed above our vulnerable raft as we skimmed across her bow or that we'd capsize as we slipped sideways in the giant swells at her stern.

Leaving the protection of Vancouver Island we headed out into the open waters of Queen Charlotte Sound. In the middle of the passage were several windswept islets whose trees, stripped bare by winter westerlies, reminded me of our luck that the wind had abated and our crossing was quite calm. Tiny phalaropes from the Arctic were bobbing up and down in hundreds on the gray-green calm sea. The flocks parted to each side as we drove toward them. How expertly they scooted and twisted and changed direction!

Our filming of the phalaropes was cut short by David's realization that we needed gas. According to what we were told in Sointula and our map, Allison Harbor on the mainland was the nearest floating supply camp.

At the dock in Allison Harbor beside a sixty-seven-foot luxury cruiser, a man all smiles was waiting to welcome us. I threw him the bowline and clambered onto the float. But he wasn't the gas man. He was an American from Seattle, a Mr. Burns who had been cruising the inlets all summer. He and his wife lived all year round in their dream boat, the *Caribou*, which had formerly been a sales boat that toured the Pacific Northwest selling Caribou-brand clothing. They were most hospitable and invited us to a dinner of roast pork and pie made with freshly picked blueberries, and an overnight stay in their guest bedroom. It was not to be, of course.

Mr. and Mrs. Burns were the only ones in Allison Harbor. The fish camp, post office, store, and gas sheds had all been abandoned the previous winter, as evidenced by a calendar on the wall of the post office. David, whose avocation in life is that of a junk-shop dealer, investigated all the paraphernalia in the deserted store. He finally salvaged a bag of nipples for himself and presented me with a battered egg beater. Mr. Burns sold us some of his own gas but we regretfully declined his kind offer for some luxury living: with another two hours of light, David wouldn't waste it on roast pork and blueberry pie.

We headed out to the open sea past Cape Caution. David wanted to spend the night on Egg Island, to inquire about nesting seabirds from the lighthouse keepers.

And then once again it rained. We buffeted our bows into each drenching wave and peered through the gloom of the encroaching dusk. Egg Island loomed closer, its shadowy outlines becoming clearer till we could see a rickety stepladder leading up to the light but with no human habitation in view. We circumnavigated the island until we came across three houses on the other side. The steep slopes of the rugged rock forbade entry in the driving rain. A winch beside a narrow channel showed how the lighthouse was serviced but no one came down from the houses to wave or winch us upward.

Our gas supply was still inadequate. As we couldn't attract any attention at the lighthouse we drove over to a nearby troller, the *Crystal Isle*, to inquire about the nearest source of fuel. We tied up alongside and were immediately invited aboard to get dry beside the galley stove over a cup of coffee.

It always fascinates me how one meets people with the same interests, even in pouring rain in the middle of the ocean. The skipper of the *Crystal Isle*, Al Delard, had recently presented the Vancouver Aquarium with a sea-lion pup. Al was very much interested to hear about our own Sam, and suggested we accompany him to Jones Cove Fish Camp, the nearest source of gas, then instructed his deckhand to tow our Canova behind the *Crystal Isle*. As it was nearly dark we thankfully accepted.

During our second cup of coffee in the warm galley, Bill the deckhand rushed in from the foredeck shouting, "Your rubber boat is loose. I guess I didn't tie it up well enough."

We rushed outside on deck to see a very small and lonely Canova fading off into the distance, a dark dot on the horizon. Al turned the troller around and we eventually regained our sole means of transportation.

By midnight we were anchored in Jones Cove. We exchanged a pound of butter for a share in Al's chicken dinner, but declined his offer of a bed on the galley floor. David and I didn't share the 4 A.M. rising habit of the ardent fisherman.

We crossed the cove to the Frances Millard Fish Camp, a well-illuminated float which housed a fish packer, to collect the fisherman's catch, a store to supply essential articles, water and ice, oil and gas facilities, and a couple of rooms for cooking, sleeping and office work.

The dock was slippery with water and salmon juice. Mountains of salmon were being dumped from the collecting fish boats onto the packer for the journey to the canneries in Vancouver. The boys working at the fish camp were all friendly university students who offered us their bunks but we watched the round-the-clock activities until 2 A.M., then laid our sleeping bags on the dirty floor beside the oil stove in the messroom. Cruising and tenting in an open boat is no fun in constant rain.

The night was hot and sleepless. David didn't think I ought to undress in a men's fish camp, so I snuggled down fully clothed in my Arctic down sleeping bag under electric light beside the oil stove and simmered. All through the night the fishermen and packers were walking over and around us to the stove, sink, and coffee table in ceaseless activity. By the time the first lot of fishermen had their fish packed, their boats iced, watered, and supplied, the next lot were coming in with their day's catch.

At dawn David breakfasted on four large Danish pastries from the store while I fried a sockeye. Outside the skies remained gray and threatening, the sea calm. About forty boats were tied up on each side of the net dock. Japanese net menders were working busily to repair a large hole about twenty-five feet long, made when a basking shark had rolled and twisted in their net.

We left Jones Cove for Bella Bella. I slept on the bow during a brief interval of intermittent sunshine as we continued our route northward along Fitzhugh Sound. David was less fortunate when we changed places. His two hundred pounds weighed down the bow and sent gallons of water down his neck. But the sea was quiet as we threaded through the many islands of the Inside Passage, passed the busy fishing village and salmon cannery of Namu, rounded Denny Island, and pulled in at the government wharf at Bella Bella.

There are two Bella Bellas: the white settlement of half a dozen houses around the B.C. Packers' store and warehouse on Denny Island, where we planned to stay; and the large Indian village of New Bella Bella on Campbell Island, two miles across the channel.

We tied up the boat, then walked to the store along the boardwalk to meet Sunny and Pat Dickson and their two delightful children, Lorilee and Melanie. The Dicksons had escaped from Vancouver to the purer air of Bella Bella about six years before. With B.C. Packers providing a house and all utilities, they had been able to save much of their combined salaries and at the same time raise their family close to the wilderness. They appreciated life in Bella Bella so much that only in an emergency did they visit Vancouver. Pat's parents even emigrated from England to live next door to them in Bella Bella.

The beauty of these northern islands and inlets developed in Sunny an increased awareness of the wild and its wildlife. A dedicated conservationist and self-taught naturalist, Sunny took long hikes into the forest behind the little rim of civilization close to the water by the store. Pat too liked animals "as long as they belong to other people."

Art, the store manager, suggested we sleep in the warehouse, so we unloaded the Canova and dumped our gear beside a pile of cedar-wood coffins. Once again we had no need to pitch our tent. In fact our sleeping quarters were increasingly unusual. Life was more comfortable all the time. Art offered toilet facilities at the store and Pat volunteered her stove for cooking.

At noon and 5 P.M. each day of the year Sunny wheels a shopping cart of meat scraps from the store along the boardwalk to the end of the dock. As soon as they see him coming, the eagles begin to congregate in the nearby trees to take their turn in diving and swooping for the handouts. We were anxious to capture as many shots and sequences as possible of the flying and diving eagles, as well as film the work of Sunny Dickson as an excellent example of conservation in action.

It wasn't easy. The first morning it seemed useless to rise before 10 A.M. while the rain poured down on the warehouse roof and no sun shone on the water below the cracks in the planking of our floor. By noon we had crawled out of our sleeping bags and had our cameras on the dock readied for action. Sunny kept throwing out his chunks of meat fat and scraps and each piece was gobbled up by the gulls instead of the eagles. In the incessant downpour, the eagles were probably sitting on hidden perches, hanging out their wings to dry. For the first two days of continuous rain, Sunny, David, and I faithfully took our positions at the end of the dock, but the eagles missed their cue.

By noon on the third day the sea fog had lifted and at last the sun appeared. And with the sun came the eagles. Suddenly the air was full of them, hovering, flying, darting, swooping. One, two, and three at a time they would take off from perches in the trees behind us to soar majestically above our heads, then with wings folded close to their bodies for increased velocity, hurl themselves like javelins toward the water. A foot above the surface the great wings unfold and they frantically back-paddle. The tail spreads wide and drops vertically to scoop back the air; in a split second the head is pulled back and the bright yellow talons are thrust forward to snatch the meat. As soon as the feet strike, the huge wings unfurl and with heavy undulating strokes slowly lift the eagle and his prize from the water.

I think more film has been wasted trying to take flying

birds in mid-air than on any other aspect of photography. We always aimed to fill the screen with eagle, which meant first finding the bird with a telephoto lens in one tiny area of a huge sky, following this moving object down to the water, then clicking the shutter with the lens exactly focused at the very moment the talons strike. Following the bird with the movie camera allowed some latitude but with the stills all settings had to be exact. The bird with wings and talons outspread must be frozen sharply in the middle of the screen. We used much film and spent many dollars with only a hope that it was to some purpose. Fortunately David could guess the eagles' movements more accurately than I and the sequence of flying eagles at Bella Bella is one of the highlights of *Coast Safari*.

Sunny, as disappointed as we were that fewer eagles were attracted to his handouts at the time of our visit than any other time of the year, said he thought there were two main reasons: "First, the weather has been so dull and rainy this summer that the birds sit tight in the trees rather than soar in the skies. And second, there's a lot more food around now than at other times of the year. They scavenge on what the fishermen and canneries throw out. The big seiners come in between Thursday and Sunday to gas and grub up for the rest of the week's fishing and they feed the eagles old salmon and rock fish. As you know an eagle, being a predator, doesn't need to eat every day. Now if you came here in winter or spring I promise you up to fifty and sixty eagles waiting for my meat."

David thought of another explanation: "This is July and at this time of the year the eagles are territorial. They are raising families and haven't yet left their nesting territories. So only those nonterritorial sub-adults and the nearby resident eagles could take advantage of your offerings. Did you notice how many more immatures we saw than adults? The adult birds are localized within a mile or two of their nests but the immatures are still free to roam around. I bet you've seen more adults in the winter and the spring than you have here all summer."

"Come to think of it, you're right," Sunny said. "I guess I should keep a daily record of how many eagles I see and whether they are adults or immatures."

David was delighted to add another eagle watcher to the lighthouse keepers, game wardens, fisheries officers, fishermen, farmers, and housewives who were already keeping daily notes for him on the eagles. This information, faithfully recorded in combination with his own observations by plane, boat, and foot, was giving David a clearer picture of where and when the eagles concentrate so that he could trace the periods and routes of their migration. With Bella Bella less than fifty miles away from Klemtu, the center of our fourth study area, Sunny's observations would be most valuable.

Behind the houses just above the village of Bella Bella was a lake where Sunny said the eagles came to bathe. I was as delighted with the parklike setting as I was with the eagles drying by the shoreline. Some were standing in the water washing themselves as robins do. Others had their immense wet wings hanging heavily by their sides. All were immatures. It was an enchanting area even when our feet squelched deep into wet bouncy bog and soft sphagnum moss.

This was an ideal opportunity for banding the birds. Eagle feathers do not have the waterproofing of seabirds so they cannot get airborne when their wings are wet. After bathing they row themselves to shore and wait on the shoreline till their feathers are dry. While I filmed the action David and Sunny herded each eagle in turn into the brush, where it was easier to grab the birds. David removed his shirt and flicked it toward the eagle, which snatched it with its talons. While the eagle fought to hold on to the shirt, David slipped his hand behind the legs and held on tightly, then closed the two wings by the eagle's side. The eagle was now held harmless as a baby. While Sunny had his first really close look at the bird he so admired, David completed the banding.

In between observing, banding, and filming the eagles at Old Bella Bella, we visited and photographed several of the nearby islands.

A few hundred yards out from the dock at Old Bella Bella is a tiny flat rocky outcrop known as Grave Island. Here is a graveyard of half a dozen headstones and a cement building housing a dozen coffins and precious possessions like silver teasets, clothes, and carvings which are buried with the Indians to be used in their afterlife.

About half a mile farther out in the channel is Meadow Island, with its more modern cemetery of plastic-flower-adorned graves set in a beautiful shady glade. In an open grave site four feet deep David found a tiny emaciated fawn, per-fectly camouflaged by its immobility and spotted coat. I won-dered if it had been left there by the doe or if it had stumbled into the grave by accident. David lifted it out to ground level and we left Meadow Island. As soon as we were gone we heard the doe rustling through the trees to her offspring.

The Indian village of New Bella Bella on Campbell Island was relocated in 1897 from the old town McLaughlin on Denny Island. With a population of eleven hundred, New Bella Bella is the largest Indian village on the B.C. coast. Just as we tied our boat to the dock, a float plane landed, carrying an elderly Indian who had hurt himself in an accident at the nearby cannery of Namu. At the end of the dock a van was waiting to take him to the only hospital between Alert Bay and Prince Rupert. The hospital van is the only vehicle in the village because, as with many of the coastal communities, there are no roads but the elevated boardwalks.

The village itself is about a mile and a quarter long and consists of several rows of houses leading up from the beach, a store, a school, and the hospital. There are no house numbers because everybody knows everybody. Washing hung from every clothesline. Children and dogs played on the boardwalk. One young girl was loading a sack of sockeye salmon onto a barrow; another was carrying a bundle of green plant stems. Indian children are usually shy or diffident with strangers and it took a while to break through her reserve. When I inquired about her bundle of green plant stems she answered hesitantly, "Gaistum. You peel it, dip it in sugar, and eat it raw. It's candy."

I think gaistum is cow parsnip. The more slender end is very tender, the thicker stem is fibrous.

With cameras and lenses strung so noticeably around our necks, I felt like a tourist as we walked along the boardwalk between the long lines of houses, some painted, some unpainted, some new, some old. We were looking for the house of Johnny Hunchit, known for his excellent carvings. But with all our camera equipment around our necks it was small wonder that the doors were not opened to our knock although people were obviously inside. Many Indians have good reason to be suspicious of strangers who barge into their reserves.

Back in Old Bella Bella the floats were choked with visiting boats. Towering over all was the *Marabelle*, a huge diesel motor vessel about a hundred and seventy feet long, owned by the Canadian Hydrographic Service for summer surveying along the coast. This year the *Marabelle* was mapping the inlets of Goose Passage between Denny and Cumberland islands. A couple of the *Marabelle* divers were scouring through kelp and beer bottles fifty feet down trying to retrieve a pair of glasses for an American tourist on one of several luxurious cruisers docked overnight at Bella Bella. These boats were participating in a race for 195 boats from Pender Harbor to Juneau, Alaska.

Some of the boaters were commenting on the large numbers of black seagulls around Bella Bella, and David soon found himself addressing a group of interested tourists on the difference between gulls and eagles. While he enlarged on his favorite topic I wandered down the float to the patrol vessel, the *Surge Rock*, for a chat on sea-lion predation with the Fisheries Inspector. During the conversation, four little Indian girls we'd met before in New Bella Bella came aboard and poked their heads into the cabin. "Hi! Can we see around your boat?" one of them asked.

Away from home, the children were very different: in their home village they had been shy and reserved. It also amazed me to know that they had rowed in the darkness two miles across the channel in a dinghy much smaller than our own. Their sturdy self-reliance was somewhat humbling.

A week later we left Bella Bella to return south. With David at the motor, I dozed all the way down Fitzhugh Sound as far as Namu, where a strengthening head wind prompted us to wait over and check the weather forecast before attempting the open seas of Queen Charlotte and Millbank sounds.

Although Namu is the home of the B.C. Packers salmon cannery, it is probably best known for what was, in 1965, the world's only captive killer whale. Two commercial fishermen from Namu had accidentally snared a five-ton killer whale in a salmon net, and named him Namu, after the town. Unfortunately even though Namu was caught in Canada, the Vancouver Aquarium was not able to muster funds to buy him from the two fishermen. Then an aggressive young aquarist, Ted Griffin, owner of the Seattle Aquarium, took the risk. He amassed eight thousand dollars overnight in small bills, flew to Namu and bought the whale, then pondered how to transport it to Seattle, 450 miles to the south.

He had two tons of precut structural steel flown up to build a floating pen. Then he chartered a purse seiner, then a tug boat, and towed Namu to the Seattle waterfront. For the first 150 miles of the journey, Namu's relatives, a cow and two calves, were in constant attendance, as Baby Doll's pod had stayed with her in Sointula. As if trying to help the captive, the cow charged Namu's cage, but, warned by her own built-in sonar, she backed off.

Sightseers at stops along the route flocked to see Namu. In Seattle, Ted Griffin swam and cavorted with Namu underwater, laying waste to the myth that this "tiger of the seas" with "the appetite of a hog, the cruelty of a wolf, the courage of a bulldog, and the most terrible jaws afloat" will attack any creature in the sea.

Anxious to continue our journey south quickly to see how Baby Doll, the Sointula killer whale, was faring, we gloomily studied the overcast sky. Heading into a southwester blowing more than twenty-five miles an hour, we would be continually soaked for the next twelve hours until we could make Bull Harbor. The dozens of fish boats tied up alongside each other had all unloaded their catches the night before and were now waiting in Namu for Sunday night. Our reverie was suddenly interrupted by a friendly voice behind us.

"Well, if it's not Dave and Lyn Hancock. What are you doing in Namu?"

"Trying to get out of it," said David. "Hi, Pete. How are you?" It was Pete Wilson, the friendly foreman of the cannery.

"Do you want a lift anywhere? A few of the fishermen are leaving for Caamano and Smith inlets: fishing's been extended another forty-eight hours. But the weather is turning bad. Where do you want to go?" Pete asked.

"Bull Harbor," replied David. "We crossed the sound on the way up here but we had the advantage of a tail wind. Rather than wait around in Namu till the wind changes, I was hoping perhaps we could hitch a lift across the sound."

"Sure. The *Klatewa* is towing fifty tons of ice on the *Cape Spruce* down to Bull Harbor. They'll be leaving in ten minutes: I may be able to arrange a ride if you like."

"That'd be tremendous. Many thanks. We're all ready."

In less than fifteen minutes we were chugging out of Namu on the fish packer *Klatewa* with the Canova bouncing jauntily over each swell at our stern. With his heavy load of ice, the *Cape Spruce* wallowed behind us on the other end of a steel cable towline. While the Indian skipper steered, his two hardy sons, clad only in cotton shirts, untangled ropes on the foredeck. We accepted the cook's invitation to have supper in the warm galley. The cook was the hereditary chief in his village and I was interested to know that he had been the right-hand man of the famous missionary Dr. Darby, who had ministered in Bella Bella for fifty years.

As the *Klatewa* rolled and pitched and wallowed in every swell and the nauseating smell of diesel fumes penetrated the galley, I began to think that there might be advantages to the cold fresh air and drenching waves in the Canova. Then as I laid my head on the galley table, the only comfortable position I could find, the engine suddenly stopped. I stood up and climbed out on deck. Had it stalled? No, the steel towline had completely unrolled from the winch and the barge containing fifty tons of ice was now cast adrift and with each wave was disappearing farther from our stern. Fortunately an hour before during supper the skipper had winched our rubber boat on deck.

The first step now was to steer the *Klatewa* close enough to the *Cape Spruce* to gaff the bowline without ramming through the barge in such heavy seas. Leaning upward from the bow, Gerry, the stalwart deckhand, somehow managed to attach a rope to the barge and slowly the steel cable was drawn up from under all those gallons of plowing and turbulent sea. Gradually the *Klatewa* chugged further away from the *Cape Spruce,* increasing the length of the line lifted through the water.

I was relieved to find the rescue of the *Cape Spruce* so easy. Then the friction of nylon rope and steel cable caused first one then another rope to fray and break. One line remained. But enough cable had been retrieved from the sea to try looping it around the winch. Just as we thought it would stay in position, the skipper slacked when he should have tightened and the entire line was lost. Once again the *Cape Spruce* drifted far behind.

With the cook, deckhands, and David in the cramped space of the afterdeck, I stayed in the galley trying to maintain an upright position by the table to find a target for the movie camera through a small porthole. As the wind intensified the waves pounded higher. The skipper decided to attach a nylon rope to the barge and very slowly tow it to calmer waters in the lee of Table Island. While I bunked on a borrowed bed the boys worked through the night to connect the cable.

Bull Harbor, an important Department of Transport radio and weather station, is a harbor in the literal sense of the word. Situated at the head of a narrow inlet on Hope Island and almost completely enclosed by land, it is a popular summer anchorage for all kinds of boats. With the barge safely docked, we thanked the skipper and stocked up on gas and supplies, filmed some fish packing, and then left Bull Harbor to surf over the Pacific swells toward Pine Island light station.

For David the attraction of Pine Island is its rhinoceros auklet colony. These nocturnal seabirds had been discovered nesting on the island in 1909 but very few observations or records had been made since then. With a westerly wind behind us we skimmed quickly over the waves to reach the steep, rocky island. Far different from our reception at Egg Island, a man and a woman descended immediately from the station and climbed down a steep ladder to meet us, and we felt welcome at once. The man called down from about fifteen feet above: "Let Lyn climb the ladder. I'll come down and help you with the boat. We can winch the whole lot up to the top."

How did he know my name? While David kept the Canova from crashing onto the rocks, I climbed up the ladder to the first flat piece of ground, where the woman greeted me with open arms: "I'm Vivian Emrich. Welcome to Pine Island, I've seen you and Dave on TV so much I feel I know you. Do you know, you're the first woman I've talked to in a year!"

I was overwhelmed. Within five minutes the Canova was winched up over fifty feet of rocks to the safety of the terrace beside the engine room and we were walking along the boardwalk past the helicopter pad, radio and storerooms, the engine room, the light itself, two immaculately painted red-and-white houses, a flower garden, a farmyard, and an abundant vegetable garden. What a paradise to find in such a remote place at the entrance of Queen Charlotte Strait—an even better seabird study area than Destruction Island!

As we reached the door of the third immaculately painted red-and-white house, Vivian said, "If I'd known you were coming, I . . ." and I finished with, "Would have made a cake?"

"No, I did *that!* I'd have swept the floor," she said, laughing.

Inside on the kitchen table was a fabulously high chocolate chiffon cake made to celebrate Ralph's birthday.

I warmed to the Emrichs immediately. They had been watching our tiny boat through their binoculars almost since we had left Bull Harbor, and as we neared Pine Island they recognized us. Ralph was a big, powerful man exuding a feeling of calm and strength in an emergency. Vivian, small and dainty, bounced around enthusiastically, her frequent smiles demonstrating her love of life and living. They were both middle-aged with grown children. I asked them when they had taken up lighthouse living.

"Viv and I first lived on a farm in Alberta, then we moved to B.C., where we started an oil-distribution business in Kamloops. About six years ago, when the family had all left home, we decided to give lighthouse keeping a try."

"And now we wouldn't do anything else," Vivian put in. "Ralph, tell them about your interview."

Ralph leaned back and puffed on his pipe before taking up the tale. "Well, I filled out all the application forms and then reported for an interview. One of the worst questions was how many tides in a day: I'd lived all my life on the prairies and never seen the sea! I came out feeling sure I'd never be accepted."

"He needn't have worried," Vivian said. "He'd always been involved with mechanical or electrical things, and that's the main part of the work. The rest you learn by experience as a junior lighthouse keeper. After Ralph was accepted we were given a list of fifteen light stations to choose from: I got out a road map, and do you know, we couldn't find a single one!"

We all laughed. I was curious to know what they liked

about life at a lighthouse, especially Pine, which is one of the three most isolated in the province.

"Getting away from the cities," Vivian answered promptly. "Where you stand for the red light and dash for the green . . . Here you have peace, quiet, clean air, time to grow your own vegetables and have a hobby. Ralph's doing woodwork; I'm tumbling stones. And here the only time you get a cold is when you go out or someone comes for a visit. You can go for a walk, get sopping wet, fall in a pool, shiver your way home, and never get a cold."

"I bet there were some disadvantages for the old-time light keepers," I commented.

Vivian nodded. "I guess it's the same old thing about slow mail. We still have that problem, especially when we have to wait a month to see how someone is, or whether it's a boy or girl. In the early days, the keepers had to row over to Susharti Post Office, miles away, for the mail and supplies. One light keeper used to send his wife: she'd row over one day and return the next. In the days before the deep freeze they'd buy a roast and cook it a little bit more every day to make it last as long as possible. Keeping meat was a problem.

"You depend on the radio for the news but it doesn't keep you up to date on the hemlines, as you find out when you leave on holiday." As we laughed, the reverberating boom of the foghorn shattered the peace and stillness of Pine.

"Well, you'd hardly believe it," Vivian said, smiling, "but I think that sound is beautiful. I sleep like a baby when old Maria is singing at night."

Ralph brought up one disadvantage of lighthouse life, which hadn't concerned them personally since their children were grown when he started his new career: "If you have school-age children who want to learn, the correspondence lessons are wonderful. But if your kid would 'druther go fishing' you've got a problem."

The Emrichs had been keen eagle observers for David over the last few years, not only on Pine but on Entrance Island, their previous posting, and I knew he was keen to talk about eagles.

"How are the eagles doing around here? I believe you have reported four nests on Pine. Rather a lot for such a small island, I would think."

"We found four nests on the island but only two pairs of adults that I think were raising young," Vivian replied. "We couldn't send you any data of nesting success because the trees are too tall to see into from the ground. This year I've seen the adults but no young. I wish I had a movie camera when the Hope Island eagle comes over to Pine. They get into a fight, lock their bills together, then fall through the air like a child's whirligig, then as they get close to the water they let go and the Hope Island bird returns to his island and the Pine Island bird returns to his nest to preen his feathers and tell his wife about it."

"Well, Vivian, most of the published information describes that behavior as the mating flight," David said. "I'm inclined to think your interpretation may be right: I've seen them mate, not in the air but on branches, and in captivity in very small pens. I think the whirligig action is either the courtship that precedes mating or aerial combat between eagles of opposing territories."

"You remember the Browns, don't you, Dave?" Vivian asked suddenly.

"You bet! Pen Brown on Pine and you on Entrance were a couple of my best eagle watchers," replied David.

"The Browns spent their honeymoon on Pine. The captain

of the lighthouse tender that delivered them to the island wagered they wouldn't last three months but they stayed ten years."

Vivian suddenly got up from the table. "I'm sure Dave and Lyn are dying to see the island, and what better way than to use the trails the Brown family cleared to all the scenic spots."

"Trails?" I echoed. "You mean we don't have to belly along the ground through the salal?"

Vivian laughed. "You can thank the Browns for that. Pine is only about three-quarters of a mile long and about two hundred and seventy feet high, so with a limited space to walk, few opportunities for boating, and a growing family to exercise, the family set to work and maintained trails all around the island."

"That's great," I enthused. "But why are there so few opportunities for boating?"

Ralph was obviously a boating man: "Well, for a start, there's no place to anchor a boat and you can't take much of a boat on the aerial tramway. And weather conditions can change rapidly. I've seen it really boil down there and winds of a hundred miles an hour are commonplace. With a two-man station, as Pine used to be, there was little time to get out and back before the next shift. Talking of shifts, why don't you go for a walk now while I get some shut-eye before the next watch?"

In the vegetable garden outside we found another of David's eagle observers, Don Vigurs, who had previously been corresponding from Active Pass. He was now the relief lighthouse keeper on Pine.

"For this kind of country your garden is doing very well, isn't it, Don?" I asked.

"Well, there have been gardens for over fifty years on

Pine," Don said. "The climate is suited fine for root vegetables. One summer the Browns grew half a ton of potatoes and seven hundred pounds of carrots. But of course abundant rain brings abundant weeds and you have to work hard to keep it from being overgrown."

"Pen Brown worked hard every winter to add to the topsoil and make it richer by bringing in barrow loads of mussels deposits from the old Indian middens. He had a more extensive garden on Pine than any of us," Vivian added.

The island was a veritable parkland. We walked easily up and down the well-marked trails to both ends of the island, along the Grand Promenade, through the Bushy Valley Trail, by Wren Walk, up Puffin Hill, and down into Sleepy Hollow. Obviously deer and goats had once roamed the island, which also helped in opening it up for walking.

That night after a supper at which we gorged on organically home-grown vegetables, we all walked along a trail to the north end of the island to study the rhinos. Once again I thought how ideal it would be to study a colony with all the comforts of home within a hundred yards.

Although little has been written about this colony of rhinoceros auklets we found that it was very extensive, the burrows accessible and well spaced, without the tangle of salmonberry typical of Triangle, Carroll, and Destruction Islands. The wings of dozens of auklets whirred past us in the darkness. We caught a few of them to check on the food they were taking back to their young in the burrows. Needlefish, cod, and sandlance were most common. David grubbed into a few of the burrows to find that the young now at the end of July were about two-thirds grown. Previously the Browns had recorded that the rhinos arrived on Pine in late April and early May, then left during late August and early September.

At one time the Indians had used the auklets for food. Missionaries have reported seeing fires on the island in the 1860's and 1870's and then seeing the Indians roasting the birds after they had been attracted to the bonfires.

As for myself, I preferred steak and mushrooms, our next day's meal on the *Sir James Douglas*, the spotless red-and-white two-hundred-foot lighthouse tender that each month services each lighthouse along the coast when the weather permits.

After a busy morning working at the auklet colony we were surprised to return to find both *Sir James Douglas* and the *Ready*, the Federal search and rescue cutter, tied up together below the light and the keepers winching up gravel for their next job of building sidewalks. We were pleased and proud to be invited aboard the *Douglas* for lunch. In my usual undignified manner I clambered into their scow to be taken out to the freighter to meet Captain Gunn and Chief Officer Finch, whose smart white uniforms contrasted sharply with our grubby, burrow-digging attire. During a magnificent lunch the two officers joked about our present form of "wilderness living."

"I remembered seeing you on TV, Lyn, axing out a shelf for your tent on Solander Island a year or so ago. Come on, tell the truth: you were probably sleeping in a centrally heated bedroom in a lighthouse," Mr. Finch said, laughing.

"And remember, Tony, when they were cooking those mussels on Triangle? I bet all the time they were really eating caviar!" Captain Gunn put in.

I almost choked on my filet mignon in the middle of Queen Charlotte Strait. "I must admit this summer's trip has scarcely been what you could call roughing it. I don't think we've put up the tent more than once. We've slept in freight sheds, fish boats, net lofts, float camps, and cedar coffins. We had steak and chocolate chiffon cake for supper last night and steak again today: you'd better keep it a secret."

Back at the Emrichs we filmed the unloading of the rest of the monthly supplies and then the work of the light itself. When I commented on the large cement foundation of the helicopter pad, Ralph explained, "That used to be the engine room but in the winter of '67 it was completely washed away in a big storm we refer to as 'the washout.' "

We urged him to go on.

"The Browns were here at the time. The high tide and hundred-mile winds blowing in from the northwest had built up a series of giant waves about a quarter mile long that scooped up ground fish to cover the beaches on Hope Island. On Pine it tore out the complete engine room, flooded the basements, and in a few minutes washed away five feet of soil that had taken centuries to form."

"Was anyone hurt?" I asked.

"No. Everything was a mess and all the houses had to be rebuilt on higher ground but nobody was injured. Now the department has made it a three-man light."

I think the Emrichs were as disappointed as we to say goodbye. Few visitors stop at Pine: boats are usually anxious to cross the sound as quickly as possible, and those that do come to Pine are either lost or out of gas.

Under unaccustomed blue skies we surged southeast toward Port Hardy, the main fishing, logging, and distribution center of North Vancouver Island. The whitecapped waves looked enormous in our wake but with a tail wind, we skimmed lightly over the surface to tie up at the dock in Sointula.

The village was deserted. All the boats were out fishing. Aaro had built a more permanent pen for Baby Doll, and she had been renamed Tula, after the community. Tula was now very tame and friendly to Barrie Palo, who was the chief feeder. The whale's own family still visited the area daily. No buyer had yet been found. "Many times we have just about cut the net and let her go," said the disheartened Palo. "But we have a lot invested in torn nets, time missed fishing, and food for the whale—she'll now eat a hundred pounds of salmon a day. We'd like to get something out of it."

We could understand his predicament but we couldn't help, and so we made a final farewell to Tula and continued south along Johnstone Strait. Our goal was Mitlenatch Island at the north end of the Strait of Georgia. David wanted to feature this nature sanctuary in our film to show the desirability of establishing more of these parks, and ways of using them in recreational education.

Two days later, at dusk, we rounded the point on the eastern end of Quadra Island. There lay Mitlenatch a few miles down the channel, our destination for the night. It seemed to be only a few hundred yards away: fifteen minutes later at twenty knots we seemed no closer. The minutes passed. It was now dark and the southern tip of Quadra Island was behind us, yet Mitlenatch seemed even farther out in the Channel. Another ten minutes and the island looked many miles ahead. David kept checking his watch and looking back. We'd been traveling for almost an hour and the island still seemed further away now than when we started. It was a most peculiar and chilling feeling. This phenomenon, related perhaps to some optical illusion because of the island's shape and size, had been known to the Indians. Their name for Mitlenatch means "The island that goes away as you come closer."

It was pitch black when we eventually arrived in Mitlenatch Bay. We followed the light of a lantern to tie up at a rock where two strangers were trying to repair an outboard motor. They and their wives had been marooned on the island that day during a fishing trip and they were now concerned that their children would be worrying about them.

Fortunately we knew Errol Anderson, the park naturalist, a tall, thin blond eighteen-year-old with an interest in natural history and talents as a bird watcher. He was employed by the Parks Branch to band the young of the glaucous-winged gulls nesting on the island and to conduct guided tours along the nature trails for tourists who daily swarmed by boat to this easily accessible island.

Errol offered the whole party a pot of coffee and settled us down on the floor in his crowded cabin. We were abruptly awakened at 6 A.M. when a huge air-sea rescue Sikorsky helicopter thundered down in the central meadow of Mitlenatch to pick up the four fishermen. No sooner had the wives been loaded aboard than a Coast Guard helicopter landed for the same purpose. Within minutes two speedboats arrived to pick up the stranded boat and motor. With such service, one certainly feels safe in B.C. waters.

Mitlenatch Island is a low, dry island with a grassy meadow in the center and two rocky promontories on the shoreline. It once belonged to the pioneering Manson family, who were living there in 1892. Cattle and sheep grazed there until 1950. The island gets less rainfall and more sunshine than either the mainland coast or Vancouver Island, so it is not surprising to find cactus in full bloom in June. The provincial government in 1960 purchased the island and designated it as a nature park.

People were now arriving in their small boats to collect the oysters that at low tide thickly crammed the entire bay. We introduced ourselves to a family from North Vancouver, Will and Wendy Thompson and their children, Roy and Glenn, who willingly acted as tourists for our film sequences.

Errol led us along the nature trails to the gull and cormorant colony, where he caught and banded some of the young. In one summer, more than a thousand gulls are banded by the park naturalists. Blinds and lookout shelters are provided to allow both education for the tourists and protection for the birds.

That evening the Thompson family invited us over to their property on the golden tropic-like sands of low-slung Savary Island, seven miles to the east. Actually Savary Island is a sandbar itself about a quarter mile wide and nine miles long. The white sand, sparse trees, and the warm waters seemed incongruous among the forested mountainous islands of coastal B.C.

While Will cooked the needlefish, anchovies, abalone, and oysters I took a quick dip in the unaccustomed warm waters of this area, and David joyfully caught a five-pound salmon with Wendy's rod and reel. After a delicious seafood-and-blueberry-pie supper we lounged on the beach and watched the sun set behind Mitlenatch.

As we headed toward Miracle Beach, another nature park on the shores of Vancouver Island, we ran afoul of some thick kelp. We stopped to remove it from the propeller and the motor refused to start again despite David's strenuous efforts. After an hour of striving we rearranged the gear and poor David suffered the ignominy of rowing an awkward rubber boat several miles against the tide into Saratoga Beach, about a hundred miles northwest of Victoria.

"We may as well phone the folks to drive up here and collect us and we'll finish the trip by road," David suggested gloomily as he bent to the oars.

"Oh no," I objected from my comfortable position as an observer in the bow. "I was looking forward to starting and finishing a whole trip at Island View Beach."

Hours later we anchored the Canova in the shallows of popular Saratoga Beach. Holiday makers suntanning on the sands sauntered over to inspect our row boat and offer suggestions. They were full of questions about where we had been and what we were doing. But the only thought in our minds was to get the motor fixed. David was all for finding the nearest garage, but a couple of men who had already started tinkering wouldn't hear of it and soon several heads were bent over the boat.

Two hours later the motor was mended but one small problem remained. The sea had deserted us. The tide had ebbed and now we had to carry our gear, the boat, and the motor two hundred yards to the water. By ourselves this would have been an impossible task: by the time we had dismantled and carried each separate article of gear down to the edge of the water, the tide would have ebbed still further. But thirty or forty swimmers and sunbathers solved the problem in short order and we were launched in style from Saratoga Beach.

After one stop to check oystercatchers on Hornby Island, David once again rowed a boat that was never meant for rowing three miles to the nearest marina that had a phone booth. David's parents came to collect us and we traveled the rest of the coast by truck.

Two weeks later we read in the paper that Tula had traveled to a large Dolfinarium in Harderwijk, in Holland. She was ferried and trucked several hundred miles to Seattle, and there loaded on a United Airlines flight for New York with a stopover at the aquarium for recuperation before her transatlantic crossing. Her constant traveling companion was Dr. Van Heel, the Director of the Dolfinarium, who sponged and soothed her for many long hours as she lay in her plastic sack. Dr. Van Heel had one bad moment when a hoist truck brought

Tula in her sling from the ferry at Kelsey Bay. The braces holding the sling were too wide for the truck and a local welder had to be found to cut away part of the braces.

Hundreds of people gathered when possible along the route from Sointula, British Columbia, to Harderwijk, Holland, to watch Tula's progress. Thousands more were to see her as the first killer whale in Europe.

First it was Moby Doll, Namu, and Tula. Later it was Shamu, Skana, and Haida. Should these killer whales have been captured or not? Would the world be a better place if all animals were left in the wild and none ever kept in captivity? Thinking emotionally one might answer yes.

Man has always regarded Nature as an enemy. He has set out to subdue or destroy. Consequently, much that was wild and free is now threatened or exterminated. If man is to turn this tide by a reasoned and a rationally guided technology then he must understand the wild and its creatures, from tiny insects to great whales of the sea. Perhaps Tula can help bridge that gap and contribute to man's understanding of one such creature in the wild.

IO

South to San Miguel

By winter David had completed our first full-length feature film, *Coast Safari*. Despite our initial worries about audience reception and the thousands of dollars we owed for financing our trips, producing and advertising the film, *Coast Safari* was an instant success. One night our bank manager slipped into the theater, not wanting to identify himself in case we thought he had come only to count the gate. Next day he told us of his great surprise at the quality of the film and the warmth of the audience appreciation; he went on to announce that, our bank account having climbed to the astronomical sum of zero dollars, he would be very pleased to let us borrow a few more thousand for the following year and another film.

David has long been concerned with oil pollution on the sea and its effect on marine life. In February he had taken *Coast Safari* down to California to show the film and examine at first hand the immediate effects of the disastrous oil leakage under Platform A, six miles off Santa Barbara.

Fifty-five hundred barrels of oil had spread out over the

Santa Barbara Channel and flowed onto the beaches and into the marinas, smearing boats, rocks, and sand. Thousands of ducks, gulls, pelicans, and other seabirds were washed up on the beaches as black sticky globs, to die of exposure after their feathers had lost their waterproofing. Bird-cleaning stations were set up at the local zoo but the results could not match up to the gallant efforts of the army of volunteers. One such volunteer wrote that it would be a miracle if one-hundredth of one percent of the oiled birds fully recovered.

David returned to British Columbia. In Santa Barbara the oil continued to gush forth from unpluggable fissures at the minimum rate of fifteen hundred gallons a day. Then in March the long ribbon of floating oil reached the island of San Miguel, forty-five miles southwest of Santa Barbara. More species of seals and sea lions breed on this island than on any other island south of the Bering Sea, and March was the breeding and pupping season. According to newspaper and magazine reports, hundreds of pathetic-eyed pups were dying. It seemed logical to David that British Columbia, with one of the most beautiful coastlines on the continent, richly blessed with seabird and sea-lion colonies, should learn more about the potential hazards of offshore drilling and bulk transport of oil in freighters; he decided that his second film, *Pacific Wilderness*, should start on San Miguel.

After spending several years studying the islands and wild-life of the northern coast, we looked forward to sunny skies and warm sea breezes as we left Island View for another summer's experiences. After the winter film tour David found he was inundated with offers of volunteer help for summer field trips. Traveling with us were Bob Flather, Grant Blundell, and Errol Anderson. Bob, with master's degrees in commerce, electrical engineering, and zoology, was now hoping to expand his zoological experience from invertebrates to marine mammals. Although he had lived many years in Santa Barbara and was a keen athlete and swimmer, he had not yet visited the Channel Islands. Grant Blundell from Vancouver Island was a shy but enthusiastic and capable lad who had just completed high school and was looking forward to his first major expedition away from home. Errol Anderson had been the park naturalist on Mitlenatch the previous summer. He had further impressed David by presenting to the Vancouver Natural History Society a vivid slide-and-sound presentation of life on a seabird colony. When David asked Errol's parents how to contact him, he was slightly taken aback at Mrs. Anderson's report that Errol was living by himself in the mountains; when we asked which specific mountain they didn't know. Somehow a message must have reached him, for ten days later, hiking southward toward Island View Beach came long-haired, barefoot Errol Anderson with a pack on his back.

In another year the first day of an expedition had found David with an eighteen-inch rip in our rubber boat. This year on our first day into California just south of Mt. Shasta, I was startled to hear a loud bump behind, then a crash. I guessed it was a flat tire. Keeping an eye on the rear-vision mirror, David pulled over, then said, "Well, that's the end of this expedition."

At times like these, David is always cool, calm, and collected, no matter what he is thinking. I am usually panic-stricken and garrulous but this time I walked back to the scene of destruction in silence.

The trailer had completely torn loose from the trailer hitch. Our brand-new sixteen-foot boat, specially constructed for offshore island visiting, lay strewn over the freeway wrecked beyond repair—and uninsured.

It had slewed off the trailer and crashed into the side of

the hill. All the contents of the trailer (cages, packboards, gas tanks, boxes of food, and containers) had spilled over the road and the adjacent ditches. The sides of the trailer were all bent and hanging from the floor, the nets were broken, and Mother's precious homemade cookies (intended to brighten up camp cooking) were flung far and wide into the dirt. The real tragedy was the loss of the boat, its new paintwork still gleaming on the splintered three-ply sides. We still had our new Avon, another type of rubber boat, but it would be impossible to fit a crew of five and our gear into one rubber boat for a journey forty miles out from shore in unpredictable Santa Barbara Channel.

There was only one consolation: the trailer had swung sideways into the side of the hill and not into the oncoming traffic. The traffic behind us had been able to swerve around us and pass. A member of the highway crew stopped, shook his head at the mess, and drove David to the nearby town of Dunsmuir. Red the Wrecker promised to be out in two or three hours. Meanwhile a Highway Patrol officer stopped: he was courteous and helpful.

Two or three more Highway Patrol officers stopped to commiserate. Bob, Grant, and Errol gathered together all that could be salvaged. They found the hammers and nails and tried to put the trailer back together again. Errol salvaged cleats, galvanized fittings, and styrofoam from the wrecked boat.

Finally Red the Wrecker arrived, surveyed the situation, agreed it was a mess, and left to get his portable welding outfit. Donning welding cap and goggles, he added stronger metal to the bars of the trailer hitch and welded them to the battered trailer.

Meanwhile we sat on our salvaged pile to drink cold beer

and make sandwiches. Our truck shook and our hair blew wildly as each semi-trailer and moving van passed us. Grant and Errol read scientific papers inside the camper, Bob studied the life of a roadside puddle, and David sought to stave off sunstroke by lying in the shade of the truck door reading *The Naked Ape*. I was the only member of the party who reveled in the scorching sun; I wrote my diary on the hood of the truck and tried to tan in the strap marks on my skin.

Four hours later, Red was finished.

We filled the reconstructed trailer, abandoned the remains of our boat to a wrecking company and continued on our journey south. To recover from our long, hot wait while the trailer was reassembled, David took the first turn off the highway to find a river where we could cool our hot bodies and dispositions. We stripped and waded over the slippery stones and boulders to drink and slosh around in the refreshing, fast-moving waters of Flume River. Nothing is as thirst-quenching as clear, cold mountain water.

Southward again through typically California landscape— expansive yellow hills rolling to the horizon, plantations of dates, olives, almonds, and stands of dull-green eucalyptus trees.

After a night parked beside the beach in the picturesque seaside community of Sausalito near San Francisco, we drove along the elevated winding coast road to Monterey. In this enchanting town with its pastel Spanish-style flower-decked cottages, our main objective was to film the thriving sea otters.

Originally the sea otter lived along the entire coast, from south California north through the Aleutian Islands in the Bering Sea to the Kamchatka Peninsula and then south to the northern islands of Japan. In 1741 Vitus Bering, commanding a Russian ship, was the first white man to see these unsuspecting and almost defenseless animals. His crewmen returned to Russia with nine hundred highly prized sea-otter pelts.

But it was not until the publication of Captain Cook's diaries that the world realized the value of the furs to be found on the west coast. Fantastically high prices were paid for pelts and Russian, Spanish, French, English, and American hunters joined the relentless slaughter that saw the sea otter eliminated from many parts of the coast as early as 1820. By 1900 sea otters were so scarce that men spent months trying to obtain one or two skins. In 1911 the sea otter became an incidental addition to an international treaty signed mainly to protect the northern fur seal, another marine mammal hunted to the brink of extinction.

Fortunately, in the remote regions of Alaska a number of animals had survived, and in 1913 the United States made the Aleutian Islands into a National Wildlife Refuge. Comprising more than two hundred wild but sheltered islands, this remote habitat allowed the sea otter slowly to rebuild their numbers.

As recently as 1938, sea otters were rediscovered in a small isolated area off Bixby Creek, a few miles south of Monterey. Under strict protection, by 1964 this population had grown to 600. Conflicting reports in 1970 set the population at anywhere from 650 to 1,500 animals. This growth delights the conservationist, but causes consternation among abalone divers. True, the sea otter does eat abalone, but this is only a small part of the diet he scrounges from the ocean floor.

We stumbled eagerly out of our truck in a quiver of inner excitement. It was hard to believe that here in a residential suburb one could sit at one's dining table and see playing in the kelp beds one of the most intriguing mammals in the entire world.

Errol was disappearing down the beach with his binoculars, excited over his first sighting of a brown pelican. The rest of us clambered over the rocks to look for a suitable vantage point for our sea-otter observations. Carefully I scanned the undulating ocean surface.

"Look, Dave," I whispered. "Over there in the kelp." I was ecstatic over my first sea otter.

Quickly David scanned the area with the binoculars, then handed them back to me with a smile. Slightly less enthusiastically, I lowered the binoculars from viewing a piece of floating driftwood lodged under a strand of seaweed.

A few minutes later we simultaneously had the feeling we were being watched. Slowly our heads turned westward, and not fifty feet from us, floating vertically half out of the water, was an adult male sea otter weighing perhaps eighty or ninety pounds, lying on its back and staring at us with eyes that sparkled not with fear but with curiosity. Before we could concentrate on cameras it had rolled head over tail in a ball and disappeared. Anxiously we crouched in the rocks and waited for its reappearance.

Otters can dive up to 120 feet and stay submerged for as long as five minutes, although most dives are less than one minute. About a minute and a half later the otter's round straw-colored, bewhiskered head popped to the surface, tiny pointed ears and black beady eyes giving it that "old man of the sea" look. One paw held what looked like a mussel, the other a flat stone. Resting the stone on his chest, the otter held one end of the mussel in each paw. Crack! The sound of the mussel being struck a number of times against the stone was clearly audible above the swishing surf. Very few animals are capable of using tools. The otter can carry a stone around for short periods in its "pocket"—a loose fold of skin near its armpits, in which it can stow choice items of shellfish, then at its leisure eat its dinner still floating on its back.

The sea otter is ever restless when awake. If it isn't sleeping floating belly up on its back, forepaws resting on its nose, flattened tail acting as a rudder, it is grooming its fine silky coat. Over and over it rolls like a porpoise, blowing air bubbles under its fur and preening at the same time to extract the dirt. Although its front flippers are short and stubby they can reach every part of its fur coat because the otter's body is so flexible inside the loose skin.

Like our pet fur seal Sam, the sea otter reached forward with one hind foot to scratch its smooth streamlined under surface, then lolled on the water, resting its forepaws on its nose. Its foot-long, slightly flat tail acted as a rudder. Occasionally it tossed and caught bits of kelp in its paws. At that point another stem of kelp got a double stare from me. In fact, the harder I stared, the easier it seemed to confuse a sea otter with a kelp frond. To make identification more difficult, otters are seldom seen outside a kelp bed, and often they drape kelp over their bodies to anchor themselves against drifting out to sea while sleeping.

We spent the afternoon filming any sea otter that came within camera range. But San Miguel was our main objective, so we continued south that evening.

At the fishing village of Morro Bay David looked for the cliff nest of the pair of peregrine falcons which had raised young there the previous year, while the rest of us ate fish and chips and watched the brown pelicans feeding in the picturesque harbor.

At last we arrived at Santa Barbara, our jumping-off point for the Channel Islands. I like Santa Barbara, particularly its spacious, tree-lined streets, the Spanish-type buildings, the white sandy beaches at its front door and the mountains at its back. Despite the Union Oil Company's mammoth cleanup job, effects of the oil spill were still visible in the blackened boulders, oil-smeared hulls of pleasure boats, and an oily coating on the soles of our shoes.

One can clean up the beaches but not as effectively can one clean up the minds and tempers of the various groups of people who months after the spill are still very much concerned with the oil situation. I found that anyone intending to visit the Island of San Miguel was viewed with suspicion. Several groups had visited the island already to investigate the effects of oil pollution on the breeding colonies. Waldo Abbott, the Curator of Mammals and Ornithology at the Santa Barbara Museum of Natural History, who surveyed the island on May 19 about eight weeks after the slick washed ashore, reported:

> I did notice a major portion of the northwest pupping beach to be heavily contaminated with crude oil. My estimates of the beach were that two hundred yards in length and seventy-five to one hundred feet wide contained heavy amounts of oil mixed with debris. In this area I counted twenty-eight dead oil-soaked pups. I also examined one or two of these carcasses and found oil in the mouth.
>
> Pups had been born in oiled areas. Those born above the oil were suckling on the oil-soaked mammary glands and fur of mothers that had crawled through oil to get to their pups. Oil contamination certainly contributed to the deaths of some of the hundreds of sea lion pups I observed on May 19.

A week later a *Life* Magazine team, accompanied by a group of concerned conservationists from the University of Santa Barbara Museum of Natural History and the Santa Barbara *News Press*, arrived to investigate. One June 13 *Life* Magazine published a report on this trip entitled "The Iridescent Gift of Death," which caused a great furor among the oil companies and certain government agencies. It was labeled "lies" and "emotional distortion of the truth." On July 3 when we arrived in Santa Barbara the headlines of the daily paper read, "Department of the Interior reports . . . NO SEALS DIED OF OIL." Other sources suggested no oil at all had contaminated the beaches.

The situation was complicated and tense.

But for our wrecked boat we would have started immediately from Port Hueneme for the island of San Miguel. As it was, with only one small rubber Avon and five in our crew it was necessary to inquire about alternative methods of transport. This proved to be frustrating. Not only did we arrive on the eve of the American Independence Day holidays, when the fishermen, abalone divers, and yachtsmen preferred to stay on shore for the celebrations, but the remoteness of San Miguel, some forty-five miles off the California coast, also meant we were at the mercies of the open Pacific. None of the fishermen we met were willing to cross the Santa Barbara Channel because of the area's history of unpredictable storms and sudden high winds. Most of the commercial fishermen had left for their summer fishing bases at Santa Catalina or Mexico. Some abalone divers said they might take us when the weather improved. But, as no one had any way of forecasting the weather out at San Miguel unless they took the plunge and went there, I could foresee a really long wait in Santa Barbara.

Also complicating matters were the conflicts between oil companies, the National Parks Branch and the California Fish and Game Branch on one hand, and the conservationists on the other. Normally the Navy, which owns San Miguel for overhead missile practice, forbids access to the island. However, after *Life*'s intrusion the Navy has issued a public statement to the effect that, had the magazine asked, it would have been pleased to give official permission to any legitimate news media. As we were preparing a program for Canadian television as well as studying the marine mammals of the island, David regarded the Navy's statement as an ideal opportunity to ask them to solve our transportation problems.

After a little stuttering, the Navy promised to take us to San Miguel in a helicopter on Monday. At the same time the National Parks Branch also consented to take us to the island on their regular patrol visit in five more days.

Monday came and went. The helicopter had engine trouble. Another date was set for Friday. On Friday the chopper had to be used in a missile-retrieval program. The Navy seemed more willing to make agreements than David to await their consummation.

Meanwhile the National Parks Branch found that in five days their patrol boat was going somewhere else. They invited David and me to join the *Cougar* for a patrol around the two Channel Islands of Anacapa and Santa Barbara. We were to accompany a university graduate student, Daniel O'Dell, who was studying the seal and sea-lion populations on these two islands.

All the Channel Islands were once connected to the mainland of southern California. Millions of years ago in an era of widespread upheaval, great land masses had intermittently risen above the ocean, later to sink slowly beneath the waves. In a period of general subsidence, a great land mass had eventually submerged, leaving eight small mountaintops protruding above the Pacific Ocean—the present Channel Islands.

Leaving Bob to entertain Errol and Grant in his home city of Santa Barbara, David and I left the town of Oxnard in the *Cougar* and rolled out to sea under overcast skies.

Thirty-eight miles from the mainland opposite the port of Los Angeles is the island of Santa Barbara, where we joined forty pleasure boats in a sheltered cove at the base of almost vertical cliffs. For about seven hours we accompanied Daniel on a pleasant stroll around the high edge of the gently rolling island. The ground was carpeted with the delicate yellow, blue, red, and white flowers of the ice plant. I was specially interested in the shriveled remains of the giant coreopsis plant, a long-lived sunflower which occasionally grows to a height of eight feet. When in the full bloom of spring, the golden yellow blossoms are visible for ten miles or more. David counted many rabbits, horned larks, meadow larks, and several sparrow hawks. Daniel took a census of the rookeries of California sea lions and elephant seals that herded together on the rocky beaches five hundred feet below.

I was disappointed at the lack of nesting seabirds. We knew that the brown pelican used to nest on Santa Barbara Island, but according to Daniel, no young had been produced in the past two years. In fact almost no brown pelicans had been able to produce young in all California. Investigations have shown that the shells of their eggs are too thin, the common symptom of pesticide poisoning. It seems that the California brown pelican is following the same route as the brown pelican of Louisiana and Texas.

Because Daniel needed to collect live mice for his laboratory studies back in Los Angeles we stayed overnight on the *Cougar*.

There was a decided feeling on our part that our host, the National Parks ranger, knowing of our interest in San Miguel, was making a determined effort to gloss over oil damage. He kept pointing out that the news media tend to exaggerate, that oil spills had always occurred naturally in the Santa Barbara

Channel, and perhaps the general public was unaware of the findings of science. He was quite certain that dying sea lions and whales, oil-covered and washed up on beaches, had nothing or little to do with oil spills. As an example he told us of recent autopsies by the government pathologists on oiled seabirds.

"You know," he said politely, "our pathologists have thoroughly dissected and studied dozens of oil-soaked seabirds in the lab. As yet they have found no link between the oil and the birds' deaths."

David interpreted this as either a sign of complete ignorance or evasion in trying to cover up one aspect of the catastrophe. Had the pathologists honestly presented a scientific judgment that was later intentionally misinterpreted by oil interests? Or was an untruthful statement presented in the first place?

It appears that because little or no oil was found in a bird's intestinal tract it was concluded that oil had not killed the bird. The smallest knowledge of seabirds makes such a conclusion ridiculous. It takes only one drop of oil to kill a seabird: with one spot of oil the feathers lose their waterproofing and clump together; sea water penetrates to the skin. Within a few hours or days, depending on the amount of skin area exposed to cold water, the bird dies through loss of body heat. Leaving the water doesn't help because cold water against a body drying in the wind causes an even more rapid loss of heat. The bird washes up on the beaches only as it nears death.

Next day we circled beautiful Anacapa Island with its craggy headlands, caves, rock bridges, and offshore pillars, dramatic results of continuous attack by wind and water.

Back in Santa Barbara we found that the next patrol of the *Cougar* to San Miguel was indefinite but that if we did find our own way there the Parks Branch would be very pleased to take us off the island. We rejoined Bob, Errol, and Grant on the promenade. They had still been unable to persuade any fishermen or abalone divers to leave the shelter of the harbor for the uncertainties of the channel. While we were discussing the possibilities of using the Avon as a ferry, three young abalone divers strolled along the promenade.

"Are you the guys who wanna go to San Miguel?" drawled the tallest one.

We looked up, all interest.

He continued. "Never been there myself. Can't tell about the weather from here. Have to go out and see. Guess we'll take you if you can leave right now. We're all gassed up. Name your price."

David was on his feet in a moment. He left word of our intentions with the Navy and the Parks Branch. In less than ten minutes we loaded our gear aboard a rather unprepossessing homemade twenty-six-foot skiff. Having been told so much about the unpredictable storms in the channel, I wore my heavy-duty emergency clothes—high boots, sweaters, and rain slicker.

They weren't needed. A mile out from Santa Barbara the sun blazed suddenly through the fog. Its dazzling light shimmered on the surprisingly calm waters of the channel. Porpoises played in our wake. Shearwaters and Xanthes murrelets flew and dived around our boat. A big bull California sea lion conveniently caught a huge flounder and ate it before our eyes. A massive eighty- to ninety-foot blue whale fed for half an hour in wide circles around our boat as we headed westward into the Pacific.

Six hours later we said goodbye to the adventurous abalone divers and threaded our way in the Avon through the many offshore rocks and reefs of San Miguel toward a small, partially protected bay.

The only residents of the island, Bob and Marty De Long, had seen us and were running over the sand dunes toward the beach. They must have wondered whether we were another group of probing writers or zealous conservationists intent on further embarrassing the government agencies that had given the oil drillers their first foothold, or just curious tourists about to disturb the nesting seabirds and the breeding colonies of seals and sea lions.

I hoped they would help to get us safely through the surf line. For some minutes we sat in the calm waters watching the waves as they crashed onto the beach. David planned to steer the boat into shore just behind a wave and at the last moment pull the motor from the water. I was ready to jump overboard and run up the beach with the long bow line. Bob would be there to grab one side of the boat and David the other. Then before the next six-foot wave could crash down on us we would be running up the beach with the boat.

All worked according to plan except for me. Not realizing the boat had such a fast forward speed I hit the beach head first, my legs folded up underneath me, and I landed ignominiously face down in the retreating surf. Bob was right there to pick me out of the water and rescue the cameras slung around my neck.

Despite my unladylike intrusion, Bob and Marty were delighted that we were not only biologists but that we had a pet fur seal of our own. David and I were invited to sleep on the floor of their cabin while Bob, Errol, and Grant were given the freedom of the eight-foot-square seal observation blind. Needless to say, the boys slept on the sand dunes.

Young, wiry, olive-skinned Bob De Long is a biologist with the United States Fish and Wildlife Service. He and Marty had come to San Miguel on their honeymoon to study fur seals. I liked Marty at once: she had the same sense of adventure and discovery on these coastal islands as I did, and we each shared our husbands with fur seals. Marty talked about her days on San Miguel with enthusiasm: "I love this island. If you read its history you'll find people are always writing that it is barren and lonely. But for me San Miguel is filled with life."

Bob told us about other people who loved San Miguel. "There's a memorial to Juan Cabrillo at the other end of the island. He came in 1543 and liked it so much that he returned during the winter and died here the following year from blood poisoning as a result of a broken leg. Then you should meet Mrs. Herbert Lester, who now lives in the hills above Santa Barbara. The Lesters came to live on San Miguel as newlyweds in 1928. Herbert Lester—he died and Mrs. Lester is a widow now—acted as caretaker for a friend who was using the island as a sheep ranch."

Here I interrupted. "A sheep ranch? San Miguel seems to be covered with sand dunes, ice plant, and shells. There doesn't seem much for sheep to graze on."

"Several ranchers before the Lesters leased the island for sheep," Bob told us. "The original brush vegetation has been overbrowsed and killed, but it did exist."

Next day the early-morning fog lifted the curtain on the rich drama of life San Miguel had to offer. We joined Bob, Errol, and Grant at the seal-observation shack, where we had piled our gear, and breakfasted among the sand dunes. Then, slung about with cameras, we set off to explore the island.

Bob De Long was already at his telescope in the blind above

the fur-seal colony. The story of the fur seal on San Miguel is a modern-day biological oddity. The valuable pelts not only of the sea otter but also of the northern fur seal brought the first explorers and exploiters to the western coast of North America. These great herds of fur seals were so rapidly slaughtered that it never was determined if this species bred anywhere but on the islands in the Bering Sea.

Then in 1965 a fisherman reported fur seals on San Miguel. Three years later when a biologist investigated these rumors he surprised the zoological world by verifying the colony's existence—one bull, a dozen females and pups had hauled out on the hot white sands of San Miguel so unlike the cold wind-ravaged rocks of the Bering Sea over two thousand miles to the north.

Bob De Long was camped here for six months to begin a long-term study that was planned to follow the expansion of these pioneering animals. The first pups observed were born on the island in the summer of 1968. At that time two tagged females were caught. Interestingly, one of the females had been tagged six years earlier in Alaska. The other had also been tagged as a pup, but on the nearby Russian rookery of the Komandorskiye Islands.

The little colony of over a hundred fur seals had their rookery about a hundred yards below Bob's tiny lookout in Adams Cove at the western end of the island, a wide expanse of sandy beach shared by much larger colonies of California sea lions and northern elephant seals. During the day either Bob or Marty was always in the shack watching the fur seals through a mounted telescope. As well as making notes of the fur seals' activities on the beach, Bob kept a careful record of the temperature of the water and the sand. Thermometers were buried down on the beach with wires leading up to the shack.

The day was hot and everywhere flippers were tossing up clouds of sand over their bodies, as a possible protection against heat or flies. As the sand grew hotter, brigades of seals flopped down to the splash zone, where the surf gave them relief. Later I sampled the water myself. San Miguel is in the path of the cold California current. Perhaps that's why the fur seals re-established themselves here.

Not having seen fur seals in the wild before, yet living with one at home, I was completely fascinated by all the "Sams" in front of me. There were about two dozen little pups as I remember our own Sam at nineteen pounds. I would have loved to walk among them as we did later with the other species but we kept at a safe distance on the sand dunes for fear of disrupting this newly established colony.

Many questions are still to be answered about fur seals. How do the females and pups navigate down from the Pribilofs in the fall? Even more remarkable, how do they find their way back to the tiny rock across twenty-four hundred miles of open ocean, out of sight of land? Because the males and females are together at only one time of the year, the fur seals have a technique for delayed implantation. (A cow's ovum can be fertilized yet she can then go for a long period before the fertilized ovum is implanted in the uterus.) How and when does this happen? Fur seals have sensitive whiskers about eighteen inches long. They can flick them back and forth with much agility. How do they use them to locate prey in the water? What is the significance of each cough and grunt in their communication? They produce a high-frequency sound underwater, similar to sounds produced by dolphins. Do they use these sounds like sonar to locate their food?

San Miguel is already the home of the Guadalupe fur seal, the northern elephant seal, the Steller sea lion, the California sea lion, and the harbor seal. Now with the re-introduction of the fur seal it has become even richer as a laboratory for studying these marine mammals.

Walking across the blossom and shell-covered dunes was relaxing compared to our usual experience of hacking through some of the thickly forested islands of the B.C. coast. Until quite recently the island had been covered by lush vegetation. As Bob had said, sheep and cattle had overbrowsed the original vegetation and now the sandy island was being further eroded by constant winds. In place of the trees and helping to stabilize the island were delicate and colorful flowers, mostly of the same ice plant family as had covered Santa Barbara Island. Scientists say that animals can be kept alive by the ice plant in the absence of water. I wondered if, in an emergency, one could drink from the white globules glistening on the undersurface of each leaf. I hoped not to have to resort to it.

Wherever we walked there were the shells of millions of land snails that had browsed the now extinct vegetation. They now formed beautiful and intricate patterns upon the sand.

Standing like white ghosts of the vegetation that once covered the islands were caliches. These calcium deposits, through some little-understood process, had formed into the shapes of plants that once lived there. One theory is that caliches originated when minerals, mainly calcium, seeped into the stems and roots of dead plants from the surrounding soil. When the winds blew the sandy soil away, left standing was the calcified plant. At the eastern end of San Miguel, eight and a half miles away, another couple, a geologist and his wife, were studying the origin of this phenomenon. Marty intended one day to walk across the island to pay them a visit.

Our normal perspective is to watch the horizon and sky for mammals and birds, but here on the wind-swept sandy stretches of San Miguel, other intriguing objects were at our feet—Indian artifacts and fossils. Once buried by time these were now exposed, with each sandstorm offering a bonanza to the archeologist, who might find everything from arrowheads to large bowls. Grant was the most observant of our crew. He picked up several perfectly shaped arrowheads and a large mixing bowl measuring a foot across, which probably dated back thousands of years. According to archeologist Dr. Philip Orr, the people of these islands are the earliest inhabitants of North America. By the carbon-dating method of aging, Dr. Orr established these Canalino people's history back thirty thousand years.

Anthropologists have drawn a number of parallels between the Canalinos and the Haidas of the Queen Charlotte Islands in British Columbia. Not only was there a similarity in the advancement of their culture and in many of their artifacts but both people were considered the most noble and most fierce in their respective areas.

Like the Haidas, the Canalinos' highly sophisticated art and their elaborate ocean-going canoes reflected the richness of the land in which they lived. The abundance of fish and sea-lion bones and sea shells in the middens or garbage dumps suggested that these people seldom worried about hunger.

Nor did we, as long as we enjoyed abalone. In one hour at low tide we could gather a hundred abalone, sometimes without getting our feet wet. However, since these measured about four to six inches across, we couldn't eat more than one or two apiece for breakfast, lunch, and dinner. Unlike our own small B.C. species which are tender and don't need much pounding, these large California ones need to be beaten strongly into submissive tenderness.

We collected the black abalone, which are considered

tougher to eat than the larger commercial red and pink varieties, but in our opinion they tasted every bit as succulent. Errol and Grant pried the abalone off the rock. Bob inserted a thin, blunt stick between the shell and the attached foot and gave a few quick jabs to separate one from the other. He cut out the soft stomach and the tough outer edges while I cut the steak into thin slices before pounding them with wood to tenderize the meat. On special occasions we dipped them into egg and bread crumbs and enjoyed a real gourmet's delight.

As we always had an excess of abalone steaks we tried feeding them to a few mangy and emaciated foxes that we daily encountered on the route to the garbage hole. To our surprise they didn't like abalone. Never had we seen them near the intertidal zone where abalone and sea-lion carcasses were in abundance. Encountering foxes on the higher interior parts of the island meant that they were feeding almost exclusively on white-footed mice and insects. On several occasions David thought he saw foxes eating vegetation. Certainly a last-choice meal for a carnivore!

The more we observed the fox, the more we realized that this delicately built animal could not be a predator on even young sea lions. Close examination of the fox's skull confirmed that his tiny teeth would never penetrate a pup's hide or grind seashells. We were observing a relic, a near extinct population of foxes that probably had done well during the island's more lush past but now were barely eking out an existence on the island's sparse vegetation and drifting sand dunes.

It is typical of remote and isolated islands that the animals are usually different from those on the adjacent mainland. The San Miguel foxes appeared half blind, squinting out of narrow slits of eyes, probably an adaptation to the constantly blowing sand and the hot, glaring sun. Because the foxes on San Miguel had never known man as an enemy they were tame enough to take food from our hands. The Lester family had loved animals. I heard that Herbert Lester would catch eagles and foxes, tame them, then on Christmas Eve let them go and find new pets.

To entice the foxes nearer for photography we threw them bacon rinds. After eating a single rind each fox rested about fifteen minutes before tackling the second one. It seemed incredible but his small stomach seemed to digest only one at a time. David held out the bacon rind, the fox stretched forward a long, skinny neck to snatch it away, while I tried to catch a close-up picture. The rind got shorter and shorter till the inevitable happened and David's finger was mistaken for the rind.

David was most intrigued to find a couple of abandoned bald-eagle nests. The one closer to our camp was a mass of twigs and sticks fourteen feet deep, overflowing from a wide crevice between two big boulders. Unfortunately no bald eagle had been known to nest there for more than twenty years. In the center of the nest was a bleached unfurred tennis ball. We wondered how many unproductive hours, days, or even weeks it had been incubated.

We had photographed the fur-seal colony through telescopic lenses from positions well hidden in the sand dunes. There was no such need to fear disrupting the gregarious elephant-seal colony. About twenty-five hundred of these amazing creatures, like overstuffed three-ton garden slugs, lay sprawled on top of each other in one black heaving mass on the sandspit. Not knowing how easy it was to approach them now that the breeding season of December to March was several months past, I had taken a whole roll of film with a telescopic lens before I realized we could walk close to them.

This was our first introduction to elephant seals, the largest of the true seal family. In the nonbreeding season, when they are hauled out for the summer fast and molt, both sexes show an incredible tameness and disregard for human beings. In the breeding season the bulls, despite their aggressive behavior toward each other, pay little attention to human beings. On the other hand the females are belligerent when a human approaches within ten feet.

David was soon in the midst of the colony to record their peculiar sounds, followed by Grant holding the other end of the tape recorder, Errol with the slide cameras, Bob with extra lenses and film, and me with the movie cameras. Occasionally a big bull left the tightly packed colony and lumbered off to the sea. Then we needed to dodge quickly, despite the entangling impediments of cameras, recorders, and extension cables. Once on the move, an elephant seal doesn't change direction very readily.

Beyond the sheer massive bulk of the elephant seal's great body, our attention was immediately drawn to the equally grotesque inflatable proboscis which gives the animal its name. Now and then two bulls reared their tremendous hulks and faced each other as if ready for combat. Usually each lifted up its long extended proboscis and waved it up and down several times, simultaneously emitting a deep-throated honking that sounded like the lowest notes of a tuba suffering from a heavy cold.

The bigger the bull, the longer his nose and the louder his voice. Scientists have two theories on how these unusual sounds are produced. One group believes that snorts from the inflated proboscis are directed into the mouth, which acts as a resonating chamber. The other theory is that expirations from the mouth are directed into the nose, which acts as the resonator.

Once heard, the sound is unforgettable. It starts as a snort then continues into a resonant almost metallic noise as if one clapped his hands sharply and rhythmically in a large metal barrel. In the breeding season the continuous vocal challenges of the bulls as they threaten subordinate males or begin a fight to maintain dominance must sound like an army retching. In the nonbreeding season the social relationships are more placid and the vocalizations less frequent.

Fights occur almost anywhere and any time, although of course more often in the breeding season. They can either be gentle pushing matches or severe earthshaking bouts that cut large gashes in the protective horny shields of each opponent's neck. Most battles now were between subdominant males trying to work their way up the ladder by challenging older bulls, or pups engaging in mock battles to prepare them for more serious fighting later.

Unlike sea lions, which will fight to hold individual females in their harem, the elephant seal doesn't actually possess a harem, although the ratio of dominant males to females is about one to thirteen. Females come and go as they please and the males are concerned with maintaining a position among the gregarious females. It takes about eight years for a bull to gain sufficient experience to hold a position among the females. But, once a male has shown his superiority in fighting, his threats and authoritative vocalizations are usually enough to discourage subordinate bulls. Some males were very old, their pink, splotched skins bearing scars from many a battle to defend their territories in the breeding season.

I liked the females best. While the bulls would sometimes weigh from five to eight thousand pounds, the females were usually less than half that weight and their liquid dark-brown eyes looked mild and cowlike.

The large eyes of the elephant seal are undoubtedly an assistance in finding food such as dogfish, shark, and ratfish in deep, dark waters.

The colony in July consisted mainly of breeding bulls, immature bulls and adult females. Where were the pups? Were they far out to sea? They were not north in British Columbia: sightings of elephant seals in B.C. have all been of adult males.

David reminded me that the breeding season for elephant seals was December to March. The pups at birth are extremely thin and weigh a mere eighty pounds. Before growing markedly in length, they become enormously fat, and by the time they are six to eight weeks old they are ludicrously obese. Their fatness almost immobilizes them and makes their movements awkward and laborious. Intensive studies of the Antarctic elephant seal have shown that the females do not feed themselves while nursing their young and may lose seven hundred pounds during the fast. Little wonder then that the pup is weaned and abandoned at two or three months of age.

Where the beach could be seen under the packed bodies of elephant seals, the sand was covered with tiny pieces of yellowish skin. Perhaps to relieve the itch of this sloughing skin, some animals had lumbered to the tide pools. Here they wallowed like hippos and blew iridescent bubbles in the green smelly, viscous muck. Others splashed about in the more aesthetic surroundings of the surf zone. Many elephant seals looked like ponderous bulky sand castles after throwing up clouds of sand with their surprisingly delicate flippers.

Sand-throwing possibly helps to relieve the burning heat or the molting skin or to brush flies away. Certainly it fills the mouth, nose, and eyes with irritating sand.

Having read little about these fascinating animals before we arrived on San Miguel, I was amazed to find that a sleeping elephant seal makes a good seat when one wants to change film or lenses. Normally they lie in such close contact to each other that small ones frequently are supported in part or entirely by their neighbors. Sometimes sea lions crawl laboriously over the bodies of the larger elephant seals, whose reaction is apparently determined by the size of the moving animal rather than by its species.

David used the back of one elephant seal as an elevated perch to get a better view of the whole herd. All of a sudden the great hulk arched its massive neck backward, flinging David and camera high into the air. As David got an upside-down view of a gaping red abyss lined with teeth, I grabbed my camera to record my husband's supreme moment—being re-gurgitated in mid-air by an annoyed bull elephant seal.

We spent the whole day with the elephant seals—photographing them from every angle in color and black and white, in movies and stills. Like the northern fur seal, the elephant seal is a species that has recovered from virtual extinction. Commercial exploitation for the animal's abundant store of blubber began as early as 1818. Then the species occupied the coast of western North America from near San Francisco along a thousand miles of coastline to Baja California. The elephant seal's habit of withdrawing slowly from a moving person makes it possible for a small group of men to herd the animals against a cliff, where they can easily be killed. By 1860 the species had become so scarce that it was no longer an important source of oil. By 1890 the species consisted of a single herd which hauled out on the exposed northwest coast of Guadalupe Island, an oceanic island lying about 220 miles off the coast of northern Baja California. It is from this remnant population that the entire species has been reconstituted. Under careful protection by the governments of Mexico and the United States, its numbers have dramatically increased from about fifty in 1892 to some thirteen thousand in 1957. The first elephant seals appeared on San Miguel in about 1938. Since then other formerly abandoned rookeries have been reoccupied.

The northwest end of San Miguel is the breeding ground of the California sea lion. On our way to the beach we appeared over a sand dune to surprise a group of nonbreeding sea lions sleeping about three-quarters of a mile from the sea. The alertness of the animal on watch was quickly transmitted through the herd and they literally galloped off in a cloud of dust. Sea lions shuffling through the desert sand out of sight of the sea! They had probably come this far inland to avoid the belligerence of the harem masters along the beach. Unlike the elephant seal, the sea lions breed during this season.

The Channel Islands are the center of distribution for the California sea lion although they do breed from Baja California to the islands off San Francisco, the former range of the elephant seal. In winter they move northward as far as southern British Columbia. Few Victorians realize that this popular sea lion of the zoos and circuses is commonly seen about February hauled out at the Race Rocks light station near the entrance to their city. And strangely enough an isolated population of California sea lions lives in the Sea of Japan.

The bachelor bulls lived among the red sandy cliffs above the pebbled breeding beaches. Surreptitiously David and I crept up to one old decrepit bull which, unable through physical strength to hold a harem, now had taken up solitary bachelorhood. While David with the still cameras wormed his way forward as close as possible to this grand old bachelor, I followed his progress through the lens of the movie camera. As the two heads gradually coalesced in the frame I began to fear for David's safety. Surely he was being too foolhardy! When the camera was within a few inches of the bull's long whiskers he jerked his head around to face us. No wonder we could get so near: he was blind in one eye, his face twisted and deformed and his jaw broken. The pronounced hump on his forehead showed he was an adult male and easily distinguished him from the other species of sea lions.

We moved on down to the water's edge: pups were huddled in their hundreds on the rocks, their mothers were swimming and resting in the adjacent waters, and the big bulls were lolling, necks outstretched toward the sky and eyes closing in the sun. As we cautiously approached, the pups paid· little attention, the females moved further away, but the bulls kept one eye open and directed toward us.

As with the fur seal, the breeding behavior of the California sea lion is highly organized. Early in the season there's a competition among the males to determine who will hold the best territories. Strength, belligerence, and bluff are the winning qualities. Deep neck gashes attested to the ferocious fights that must have occurred early in the breeding season. Now in July altercations had diminished to mere pushing and bluffing.

The females come ashore in May to pup and be mated. By ten days of age the pups start to explore the tide pools and play in the surf but they will not be weaned for at least another five or six months.

"Look, Dave, there's a pup calling for its mother," called Errol.

"Good man! We'll wait to see which female responds. I want to film the reactions between them," answered David as he set up the cameras.

We were lucky! One of the females socializing with the others in a tide pool came over to smell the calling pup. Realizing that it wasn't hers, she showed no more concern. Another mother approached, recognized it as her own, and like a dog or cat, she very carefully picked up her pup by the neck folds and swam away to another rocky outcropping further down the beach.

With camera gear trailing after us, David and I hurried down the pebbled beach, intending to film what she did with her offspring from as close an angle as possible. No sooner was the tripod in place than she gently deposited the pup amidst some others on another rock. Sea lions have excellent sight and a sufficiently good sense of smell to identify their young out of hundreds on the beach.

Later in the afternoon, while Errol, Grant, and Bob were helping David with interaction sequences for the film, I wandered down the tide-exposed beach to concentrate on taking portraits with the still cameras. While other bulls nearby lumbered seaward at my approach, one majestic old fellow, deeply gashed by recent battles to hold his territory, sat defiant at the water's edge.

He was such a royal subject for photography, but how close could I encroach? At fifty feet his great bulk filled the frame as he watched me through half-closed lids. Forty feet, click; thirty-five feet, click, click. At twenty-five feet only his magnificent head was framed. And his bloodshot eyes were fully opened.

Wishing to take an even tighter closeup, I edged forward carefully until my camera rested on a shelf of rock separating us by a mere ten feet. I was amazed that my close presence had so little effect on him. My lens now centered on one blood-red, saucer-shaped, attentive eye. To get better support for my camera, I placed one foot on the other side of the rock. That did it. Half a ton of resting blubber erupted into half a ton of charging fury. With a deep growl his thick bull neck reached out like a concertina until his long canines lunged within inches of my face. I stumbled back over the slippery rocks in terrified retreat, remembering only a blurred image of canines coated with a brown scum, long, stiffly protruding whiskers, and a loud cough eclipsed by my own sudden screams. A few yards up the beach I missed my footing and fell forward onto the rocks. David ran up to pull me to my feet. I looked backward to see my pursuer, his haughty pose resumed, sitting again at the water's edge on the far side of the intervening rock.

David, as agitated as I was, laughed nervously. "Don't do that again," he scolded. "Only last week I was told of a chap in another part of California who foolishly mistook the calm tolerance of a territorial bull for tameness, and walked directly into his territory. Without warning and before he could back-track, he was grabbed by the backside, picked up, and thrown several yards. He lost such a hunk of rump that he died from loss of blood."

I had been lucky. While my feet stayed on my side of the rock boundary that marked the edge of his territory, the bull remained alert but tolerant. However, as soon as I invaded the limits of his defendable area, he attacked.

Feeling limp, I rejoined the others, leaving the bull in command of his own bailiwick.

At one end of a half-mile deserted bay we encountered a small gathering of pups. At first glance everything seemed in order. Then as we moved closer we saw blotches of oil on many of their underbellies. They had wriggled across some of the globs which the waves had deposited on the beach.

Only four small bays, perhaps not encompassing more than half a mile of beach, were inundated by oil from the Santa Barbara spill. But because nearly all of the shoreline of San Miguel's thirty-four-mile coastline is made up of precipitous rocky faces and only the extreme west end of the island contains the sandy beaches on which the seals breed and pup, the oil-slicked bays represent a large percentage of the breeding and pupping area.

We saw many dead and dried-up fetuses and several emaciated pups, some dying, some dead. I wished I could do something for them. When I stayed beside one he'd initially growl and bare his teeth but if I held my ground he'd whisper and nuzzle up to me with an almost pleading look in his eyes. Perhaps sadder than the pups that had died quickly were those that lived on the edge of the slick. Only partially oiled, they were dying a lingering, pathetic death. Patches of oil on the pup's hide can cause loss of insulation, massive irritation and burning, skin poisoning, and loss of identification through smell, with subsequent starvation.

I wanted to cuddle them, to comfort them, to try to help them. But we were too late: volunteers could have offered some relief to a few pups had they been allowed to remove them from this preserve, but help could only be minimal.

By this time some of the oil remained soft and mucky but most had been oxidized to asphalt. With jeans and boots permanently caked with oil from the black rocks, we walked slowly back toward camp, hearts heavy.

Descending a high ridge coated with ice plant, I looked up to see that two boats had anchored in the bay. The first we immediately recognized as the *Cougar*, the National Parks Patrol boat which was tentatively to return us to the mainland in a couple of days. But why the second?

Our questions were soon answered when a dozen or more people came over the hill toward us. Without delay we were told that we would be leaving within an hour on the second boat, the Union Oil Company charter boat, which had brought to San Miguel some oil-company journalists and a marine research team from the Hancock Foundation—unfortunately not these Hancocks.

Bob De Long came running out of his observation shack to meet the group and lead them over to the sea lion beaches. While David was called aside by the park ranger for a private chat, I asked the editor of the *Union Oil* Magazine why he had come to San Miguel. "We have come to refute the claims of a *Life* Magazine article recently published about the effects of oil on San Miguel," he curtly replied as he set off over the sand dunes after the main group.

We had read and reread that article several times to find some variance from the truth as we ourselves found the situation on San Miguel. The writer had mentioned that he "didn't bother to request official permission from the Navy to visit the island"—rather an undiplomatic admission for such a major magazine, and certainly both subject and style were emotional, but in the light of our experience on San Miguel we found no inaccuracies.

I think what astonished me most was the editor's arrival on San Miguel a month after publication of an article he intended to refute, the fact that his mind was made up before setting foot on the sea-lion beaches, and the very short time the investigators remained on the island.

David returned from his chat with the park ranger. He was terse and to the point. "Bob, go and get our sleeping bags from the De Longs' cabin. Errol and Grant, break camp and take the gear down to the beach as quickly as possible. The Parks Branch has asked me to screen all the film we have taken before we leave the area. They're sure touchy about the oil situation."

The boys rushed off to break camp while David and I

reeled off a dozen rolls of movie film on scenery and plant cover. Back at camp thirty minutes later, the boys were carrying the assembled equipment down to the beach. As we were taking a few last shots of Bob and Marty at the fur-seal observation shack, the scientists and journalists from the investigation party were seen streaming back from the oiled beaches.

During our stay on the island we had avoided discussion of the controversial oil situation. Our conversation was restricted to the archeology and history of San Miguel and Bob's study of the fur seals. Bob was to stay on the island until October, when the fur-seal colony migrated to sea. Marty was to leave for the mainland in another month to prepare for her teaching job in the fall.

We were very loath to leave San Miguel and these warm friends in such a rush, but without a boat of our own we had no choice. Despite our clearance to visit San Miguel it was obvious that any photographer, zoologist or not, was regarded with suspicion.

During the trip back to Santa Barbara we got acquainted with the oil-company representatives. Dr. Carleton Scott, technical manager of an oil company chemical subsidiary, was the leader of the little expedition. His aims were to take a look at San Miguel and its animals and to check a theory that the oil that had washed up on San Miguel might have come from a natural seep. The other scientists aboard were out to collect specimens for a study financed by the petroleum industry concerning the effect the oil might have had on animal and plant life in the area. The three journalists aboard intended to return to their oil-company magazines and write authoritatively from their one-hour observations. I appreciated a little more the approach of Bob Gaines, the young, likable editor of *Petroleum*

Today. He intended to concentrate less on the oil effects and more on the fascination San Miguel held for those who visited it.

San Miguel *is* fascinating—a sturdy island which we hope will continue to withstand bombardment by man and missile as it has the ravages of wind and water.

II

The Morning of Another Day

DURING SEVEN YEARS OF STUDYING, COLLECTING, AND FILMING, our home has been filled with animals. The story of their adventures would make half a dozen books. But, until I can get them written, let me give you a few scenes from the continuing drama of Hancock Hostel:

SCENE: The Hancock home and Wildlife Conservation Center in Saanichton, near Victoria, British Columbia.

TIME: A day in the life of . . .

MAIN CHARACTERS: Sam, a fur seal washed up on the beach seven years ago. Gypsy, the gibbon, orphaned when her mother rejected her and broke her arm. Bubu, the bear cub, orphaned when some loggers burned out the den and the mother ran off. Chimo, Chica, and Tom, month-old cougar kittens also rejected by their mother. Scarlet, a macaw, fifty years old or more, who loves my husband and hates me.

Pixie and Pete, two smart coatimundis, appealing creatures with long inquisitive noses. Porky, the porcupine, who has a gargantuan appetite. Pierretta, the Canada goose, who dropped in one day and won't go away. David, the husband, who comes and goes—mostly goes, doing his bit to save the wilderness world. Me, the wife, feeder, and general doer of whatever needs doing at any particular time. Supporting characters: Ducks, eagles, falcons, seabirds.

7 A.M. / Bubu, the bear cub, on the back porch, wakes me up with her penetrating cries for breakfast. She is caged at night because she has the disturbing habit of climbing the shelves in the pantry. A bear cub's crying is disconcertingly like that of a baby, and the cub requires just as much attention as a baby— perhaps more, I reflect as I lie in bed and listen to her crying for half an hour, then give up and get up.

From her perch on the TV set, Scarlet, the macaw, whispers her morning soliloquy as I dress: "Hello. How are you? What's your name? My name is Hancock." As I leave for the kitchen to prepare the bottles, she calls out a shrill "Goodbye!"

Gypsy swings into action as I pass her pen, and tries to strangle me into stopping by reaching through the wire and winding her long arms around my neck.

In the kitchen I plug in the electric kettle to boil the water for the cougar formula, sterilize the bottles, then mix in the vitamins and calcium additives with the milk. I prepare a big bottle for Bubu, medium bottles for the three cougar kittens, and a small bottle for Gypsy. While the bottles are warming I put Scarlet, the macaw, with her day's supply of sunflower seeds and water, on her perch in front of the house, where she can spend her day talking to the beachcombers and rockhounds. On the way back to the kitchen, I turn over the barn-owl and falcon eggs in the incubator and check the temperature.

8 A.M. / I feed the cougar kittens. Chimo drinks eagerly and purrs like a motorboat as he nestles into my hand and sucks on my arm. Chica needs coaxing; Tom has to be force-fed to get enough nourishment into him.

I wash Gypsy in the tub, getting washed myself in the process. I haven't housetrained her in the sense in which cats and dogs are trained, so she wears diapers. Then with Gypsy clinging to my back I pick up Bubu on the porch and try to stand upright under her onslaught on the bottle. The three of us continue to the compound, where Gypsy and Bubu play around my legs and get in the way while I try to do the other chores. Sometimes I stumble from pen to pen dragging one bear cub and supporting one gibbon.

8:15 A.M. / I cut up the fruit for Porky and the coatis, Pixie and Pete, who chitter a welcome to me as they walk tightropes upside down in their enclosure. With Bubu and Gypsy following behind me and a bucket in each hand, I feed the coatimundis, Pixie and Pete, who climb all over me as I open the door. They are the most inquisitive creatures I know and their long noses ferret into my hair, my ears, and my nose. They have an excellent knack for opening doors, but after finding them sitting on the ledges of the falcon pens one day looking down onto the falcons that they had displaced, we devised a door system which so far withstands their ingenuity. They are warm, friendly creatures but impossible to keep in the house because they are so active and curious. Last time they paid us a visit indoors they chewed right through the telephone wire and

I was cut off—literally. Porky would like attention too, but how do you pet a porcupine? Next I feed the pigeons, the ducks, and Pierretta, the Canada goose.

Recently a Japanese television network sent a team to Canada to interview David for a teach-in on pesticides to be held at Tokyo University. Part of the interview was filmed in a field behind our compound while Dave was strolling along with Bubu. Most of the sound had been taped the previous night—before the producer, the only one of the team who could speak English, had rushed off to Texas on an urgent assignment.

The two gentlemen who were left communicated mostly in bows and smiles. Imagine their surprise when out of the nearby woods ambled a goose. It was Pierretta on her morning stroll. After an opening monologue of honks from Dave, the goose joined in the conversation and waddled along home with the bear. (I am sure Dave will be labeled the Konrad Lorenz of Canada when the film is shown in Japan—The Man That the Wild Geese Follow!)

Back to the preparation room to fill a bucket with at least twenty-five pounds of herring for Sam. He shares a very large swimming pool with some seabirds and a California gull that came to us originally with a broken wing and stayed probably because he never had it so good in the wild.

About this time Bubu usually gets into some kind of scrape. Bears have a nose for trouble. She tries climbing up the wire sides of the seabird pen. Climbing up is easy and a stroll on top of the roof looking down at the other animals is fun. But how will she get back?

First she pokes her nose over the edge and looks down. Now how did the ground get to be so far away? She turns herself around hoping to get down rear first, and dangles her hind legs over the edge, keeping a strong grip on the roof with front paws. No, that's no good either. Perhaps by running up and down along the roof she'll find an easier place that's closer to the ground. By this time her cries are so pitiful I get a ladder and rescue her myself.

The birds of prey are next. I hack up chicken into pieces suitable for the size of each bird. A messy job, but I've come a long way since those first days when I would close my eyes and madly mangle away. Gypsy usually observes all these operations from her stranglehold on my neck, so I have to put her forcibly down on the ground when I go to feed the birds. Many of these species are endangered in the wild, and Dave is hoping to breed them in captivity, so I prefer not to upset them by barging into their enclosure accompanied by bears and gibbons. I collect yesterday's leftovers, renew their water, and place new food on the ledges. Baldy, the eagle with the amputated wing, kacks out her morning greeting. Most important now is cleanup time, when all the utensils and buckets are scrubbed, the floor is swept, the garbage collected, and chicken and fish are laid out to thaw for the following day.

9:30 A.M. / Bubu and Gypsy follow me back into the house. Protesting volubly, they are deposited in the pen in the living room while I skim through some housework and prepare the bottles for the cougars' second feeding. Gypsy is the most playful, swinging on the ropes and along the bars, while Bubu climbs up on the middle shelf and falls asleep oblivious of the occasional kick from Gypsy's swinging feet.

10 A.M. / With a cougar in one hand and a milk bottle in the other, I have to answer the phone. A local school group wants to come out on a field trip. Details are discussed and a date arranged.

I am just about to sit down to feed the second cougar when the phone rings again. Could David address an antipollution group the following month?

10:30 A.M. / The doorbell rings just as I am massaging the rear end of the third cougar to encourage it to make a bowel movement: mother cougar would encourage bowel movements as she groomed them. Cougar in hand, I go to the door. It is a picnicker who wants a bucket of water and the use of the phone.

11 A.M. / The mail arrives. I pick up some bottles and papers left by the previous day's visitors at the beach and deposit them in the garbage can, then go into David's workroom to sort out the mail. Scientific journals are quickly catalogued, articles of importance are marked and clipped for Dave's attention, photos are viewed and classified.

NOON: / I realize it is about time I had breakfast so I grab my favorite meal, malt bread and cheese, while I sterilize bottles and nipples.

I give Bubu and Gypsy a run around the house for exercise. They chase each other under beds, behind sofas, behind the stove, and into the bath. When Bubu gets too obstreperous, Gypsy has the advantage: she can shin up the curtains and swing from the curtain rods well out of the bear cub's reach. Bubu avenges herself by chewing at the covers of an easy chair.

With Gypsy for company, I race off downtown in the car to pick up chicken from the poultry processors, fruit and vegetables from the supermarket. I leave some room in the car for Sam's herring, which I collect from the rented locker.

3 P.M. / Gypsy and the cougars are fed again. Bubu is having her afternoon nap. I settle down to the typewriter to continue

writing the current chapter in a book. The phone rings again: a high-school student wants help with a school biology project, and I suggest she come around and use Dave's library.

5 P.M. / I leave the typewriter and open the fridge. If Dave is home, I spend about five minutes thinking about our evening meal. If he is away I cut off a slice of ham and several slices of cheese, then settle down with a cup of coffee to peruse the evening newspaper.

I clip articles of interest to Dave, then take Bubu and Gypsy down to the beach for a brief romp. Bubu usually has to be carried down to the sea, well away from the house, because when she is allowed out her first thought is to go back inside again. But, once she gets far enough away from the front door, she follows close behind me.

Back at the house, she gets another bottle and is put back in the pen with Gypsy. By now Gypsy's metabolism has slowed considerably, and she curls up on the top shelf and tries to sleep.

When she's awake, Gypsy wants the security of a mother, so her favorite position is hanging around my neck. A gibbon tightly wrapped around one's neck is somewhat of a hindrance to typing, especially when she prefers to sit on my lap rather than hang from my back and look over my shoulder.

Scarlet, the macaw, is brought in from outside the house and cracks open sunflower seeds from her perch on the TV to my right. Bubu alternately sleeps and plays in the pen to my left; the cougar kittens practice walking in their funny little uncoordinated leaps and bounds on the living-room carpet at my feet and the book goes on.

8 P.M. / The phone rings. There is an otter in Oak Bay, sick and washed up on the beach. Will Dave go and pick it up? I

race out to the compound to grab nets, gloves, and a pen. It is dark by the time I arrive and I can't find any people or any otters. I return to the typewriter.

MIDNIGHT. / Gypsy by this time has crept into my bed and is curled up between the sheets. I turn on the light and she squints up at me out of one eye to see if I'll let her stay for the night. But I know Gypsy. She will lie quiet and cuddly till daylight, then will leap into action by swinging on the curtains and pulling clothes from the hangers in the closet. I harden my heart, pull her clinging arms from my neck, and deposit her in the living-room pen, where she cries for a while, then forgets, and goes to sleep till morning.

12:10 A.M. / David arrives home on the last ferry from the mainland, where he has been working on our next wildlife film.

2:30 A.M. / The front doorbell rings. It can't be a burglar—he wouldn't announce his presence. Probably it is a couple whose car needs to be towed out of the sand dunes at the beach. No. It's a gentleman who has run into a thin and emaciated barn owl on the highway. It seems stunned. There are no bones broken. We feed it and put it in a dark box till morning.

And then it is the morning of another day.

A Note About the Author

Lyn Hancock was born in 1938 in Fremantle, Western Australia. She attended teacher's college and taught high school, specializing in speech and drama. After studying speech arts in Australia and in England at the Royal Academy and Trinity College, she ran a private academy for speech arts. In 1960 Mrs. Hancock embarked on three years of traveling, studying and teaching in Southeast Asia, Africa, Europe, and North America. On the eve of her return to Australia she met her husband, David, in Vancouver, British Columbia. Mrs. Hancock taught in Vancouver for four years before resigning to work full time with her husband in the conservation field. The Hancocks run the Wildlife Conservation Center in Saanichton, British Columbia, and together they produce films and lecture on their experiences. Mrs. Hancock is the author of numerous articles and two books.

A Note on the Type

This book was set on the Linotype in Janson,
a recutting made direct from type cast from
matrices long thought to have been made by the
Dutchman Anton Janson, who was a practicing type
founder in Leipzig during the years 1668–87.
However, it has been conclusively demonstrated
that these types are actually the work of Nicholas Kis
(1650–1702), a Hungarian, who most probably
learned his trade from the master Dutch type
founder Dirk Voskens. The type is an excellent
example of the influential and sturdy Dutch types
that prevailed in England up to the time
William Caslon developed his own incomparable
designs from them.
This book was composed, printed, and bound by
The Book Press, Brattleboro, Vermont. Typography
and binding design by Guy Fleming.